# SQL Server Big Data Clusters

## Data Virtualization, Data Lake, and AI Platform

## Second Edition

**Benjamin Weissman**
**Enrico van de Laar**

Apress®

***SQL Server Big Data Clusters: Data Virtualization, Data Lake, and AI Platform***

Benjamin Weissman
Nurnberg, Germany

Enrico van de Laar
Drachten, The Netherlands

ISBN-13 (pbk): 978-1-4842-5984-9
ISBN-13 (electronic): 978-1-4842-5985-6
https://doi.org/10.1007/978-1-4842-5985-6



Managing Director, Apress Media LLC: Welmoed Spahr
Acquisitions Editor: Jonathan Gennick
Development Editor: Laura Berendson
Coordinating Editor: Jill Balzano

Cover image designed by Freepik (www.freepik.com)

Distributed to the book trade worldwide by Springer Science+Business Media New York, 233 Spring Street, 6th Floor, New York, NY 10013. Phone 1-800-SPRINGER, fax (201) 348-4505, e-mail orders-ny@springer-sbm.com, or visit www.springeronline.com. Apress Media, LLC is a California LLC and the sole member (owner) is Springer Science + Business Media Finance Inc (SSBM Finance Inc). SSBM Finance Inc is a **Delaware** corporation.

For information on translations, please e-mail rights@apress.com, or visit http://www.apress.com/rights-permissions.

Apress titles may be purchased in bulk for academic, corporate, or promotional use. eBook versions and licenses are also available for most titles. For more information, reference our Print and eBook Bulk Sales web page at http://www.apress.com/bulk-sales.

Any source code or other supplementary material referenced by the author in this book is available to readers on GitHub via the book's product page, located at www.apress.com/9781484259849. For more detailed information, please visit http://www.apress.com/source-code.

Printed on acid-free paper

*This one is dedicated to all the ravers in the nation.*

# Table of Contents

# About the Authors

**Benjamin Weissman** is the owner and founder of Solisyon, a consulting firm based in Germany and focused on business intelligence, business analytics, data warehousing, as well as forecasting and budgeting. He is a Microsoft Data Platform MVP, the first German BimlHero, and has been working with SQL Server since SQL Server 6.5. If he's not currently working with data, he is probably traveling and exploring the world, running, or enjoying delicious food. You can find Ben on Twitter at @bweissman.

**Enrico van de Laar** has been working with data in various formats and sizes for over 15 years. He is a data and advanced analytics consultant at Dataheroes where he helps organizations get the most out of their data. Enrico is a Microsoft Data Platform MVP since 2014 and a frequent speaker at various data-related events all over the world. He writes about a wide variety of Microsoft data-related technologies on his blog at enricovandelaar.com. You can reach Enrico on Twitter at @evdlaar.

# About the Technical Reviewer

**Mohammad Darab** is a data professional with over 20 years of IT experience, 10 years of that working with SQL Server. He's a speaker, blogger, and a self-proclaimed Big Data Cluster advocate. Since the introduction of Big Data Clusters in SQL Server 2019, Mohammad has been actively advocating what he calls "the future of SQL Server" through his social media outlets and blog at `MohammadDarab.com`.

When he's not creating Big Data Cluster content, he's spending time with his wife and their three kids in their home in Virginia.

# Acknowledgments

As with every publication, a big THANK YOU goes to our families for the support they gave us during this time-consuming process!

Also, thank you very much to Mohammad for his support by reviewing this book!

We would also like to thank the Microsoft SQL Server product team for helping us out whenever we had a question or ran into situations we didn't quite understand. JRJ, Travis, Buck, Mihaela, and all the others – you rock!

Last but not least, thank you #sqlfamily – your ongoing support, feedback, and motivation is what keeps us going when it comes to exploring and talking about exciting technologies like Big Data Clusters!

# Introduction

When we first started talking about writing a book about SQL Server Big Data Clusters, it was still in one of its first iterations. We both were very excited about all the technologies included in the product and the way it could potentially change the field of data processing and analytics. Little did we know how much changes the product was going to receive while we were writing this. Ultimately this resulted in us almost rewriting the entire book on a monthly basis. While this was a massive endeavor, it also allowed us to follow, and document, everything the product went through during its development. Now that the final product has shipped, we thought it was about time to provide an updated version that reflects everything that Big Data Clusters is today; the result is the book in front of you right now!

SQL Server Big Data Clusters is an incredibly exciting new platform. As mentioned earlier, it consists of a wide variety of different technologies that make it work. Kubernetes, HDFS, Spark, and SQL Server on Linux are just some of the major players inside a Big Data Cluster. Besides all these different products combined into a single product, you can also deploy it on-premises or in the Azure cloud depending on your use case. As you can imagine, it is near impossible for a single book to discuss all these different products in depth (as a matter of fact, there are plenty of books available that do go into all the tiny details for each individual product that is part of a Big Data Cluster like Spark or SQL Server on Linux). For this reason, we have opted for a different approach for this book and will focus more on the architecture of Big Data Clusters in general and practical examples on how to leverage the different approaches on data processing and analytics Big Data Clusters offer.

With this approach, we believe that while you read this book, you will be able to understand what makes Big Data Clusters tick, what their use cases are, and how to get started with deploying, managing, and working with a Big Data Cluster. In that manner this book tries to deliver useful information that can be used for the various job roles that deal with data – from data architects that would like more information on how Big Data Clusters can serve as a centralized data hub to database administrators that want to know how to manage and deploy databases to the cluster, data scientists that want to train and operationalize machine learning models on the Big Data Cluster, and many more different roles. If you are working with data in any way, this book should have something for you to think about!

# Book Layout

We split this book into nine separate chapters that each highlight a specific area, or feature, of Big Data Clusters:

- Chapter 1: "What Are Big Data Clusters?" In this chapter we will describe a high-level overview of SQL Server Big Data Clusters and their various use cases.
- Chapter 2: "Big Data Cluster Architecture." We will go into more depth about what makes up a Big Data Cluster in this chapter, describing the various logical areas inside a Big Data Cluster and looking at how all the different parts work together.
- Chapter 3: "Deployment of Big Data Clusters." This chapter will walk you through the first steps of deploying a Big Data Cluster using an on-premises or cloud environment and describe how to connect to your cluster and finally what management options are available to manage and monitor your Big Data Cluster.
- Chapter 4: "Loading Data into Big Data Clusters." This chapter will focus on data ingression from various sources unto a Big Data Cluster.
- Chapter 5: "Querying Big Data Clusters Through T-SQL." This chapter focuses on working with external tables through PolyBase and querying your data using T-SQL statements.
- Chapter 6: "Working with Spark in Big Data Clusters." While the previous chapter focused mostly on using T-SQL to work with the data on Big Data Clusters, this chapter puts the focus on using Spark to perform data exploration and analysis.
- Chapter 7: "Machine Learning on Big Data Clusters." One of the main features of Big Data Clusters is the ability to train, score, and operationalize machine learning models inside a single platform. In this chapter we will focus on building and exploiting machine learning models through SQL Server In-Database Machine Learning Services and Spark.

- Chapter 8: "Create and Consume Big Data Cluster Apps." In the second to last chapter of this book, we are going to take a close look at how you can deploy and use custom applications through the Big Data Cluster platform. These applications can range from management tasks to providing a REST API to perform machine learning model scoring.
- Chapter 9: "Maintenance of Big Data Clusters." To finish off your Big Data Cluster experience, we'll look at what it takes to manage and maintain a Big Data Cluster.

CHAPTER 1

# What Are Big Data Clusters?

SQL Server 2019 Big Data Clusters – or just Big Data Clusters – are a new feature set within SQL Server 2019 with a broad range of functionality around data virtualization, data mart scale out, and artificial intelligence (AI).

SQL Server 2019 Big Data Clusters are only available as part of the box-product SQL Server. This is despite Microsoft's "cloud-first" strategy to release new features and functionality to Azure first and eventually roll it over to the on-premises versions later (if at all).

Major parts of Big Data Clusters run only on Linux. Let that sink in and travel back a few years in time. If somebody had told you in early 2016 that you would be able to run SQL Server on Linux, you probably would not have believed them. Then SQL Server on Linux was announced, but it was only delivering a subset of what it's "big brother" – SQL Server on Windows – actually contained. And now we have a feature that actually *requires* us to run SQL Server on Linux.

Oh, and by the way, the name is a bit misleading. Some parts of SQL Server Big Data Clusters don't really form a cluster – but more on that later.

Speaking of parts, Big Data Clusters is not a single feature but a huge *feature set* serving a whole lot of different purposes, so it is unlikely that you will be embracing every single piece of it. Depending on your role, specific parts may be more useful to you than others. Over the course of this book, we will guide you through all capabilities to allow you to pick those functions that will help you and ignore those that wouldn't add any value for you.




B. Weissman and E. van de Laar, *SQL Server Big Data Clusters*,
https://doi.org/10.1007/978-1-4842-5985-6_1

# What Is a SQL Server 2019 Big Data Cluster Really?

SQL Server 2019 Big Data Clusters are essentially a combination of SQL Server, Apache Spark, and the HDFS filesystem running in a Kubernetes environment. As mentioned before, Big Data Clusters is not a single feature. Figure 1-1 categorizes the different parts of the feature set into different groups to help you better understand what is being provided. The overall idea is, through virtualization and scale out, SQL Server 2019 becomes your data hub for all your data, even if that data is not physically sitting in SQL Server.

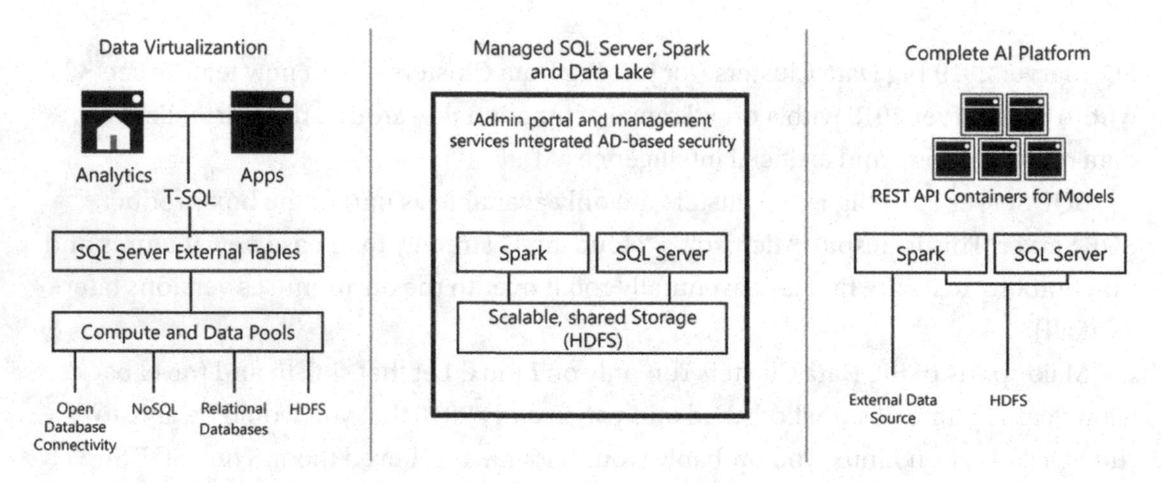


***Figure 1-1.*** *Feature overview of SQL Server 2019 Big Data Clusters*

The major aspects of Big Data Clusters are shown from left to right in Figure 1-1. You have support for data virtualization, then a managed data platform, and finally an artificial intelligence (AI) platform. Each of these aspects is described in more detail in the remainder of this chapter.

# Data Virtualization

The first feature within a SQL Server 2019 Big Data Cluster is data virtualization. Data virtualization – unlike data integration – retains your data at the source instead of duplicating it. Figure 1-2 illustrates this distinction between data integration and data virtualization. The dotted rectangles in the data virtualization target represent virtual data sources that always resolve back to a single instance of the data at the original source. In the world of Microsoft, this resolution of data to its original source is done via a SQL Server feature named PolyBase, allowing you to virtualize all or parts of your data mart.

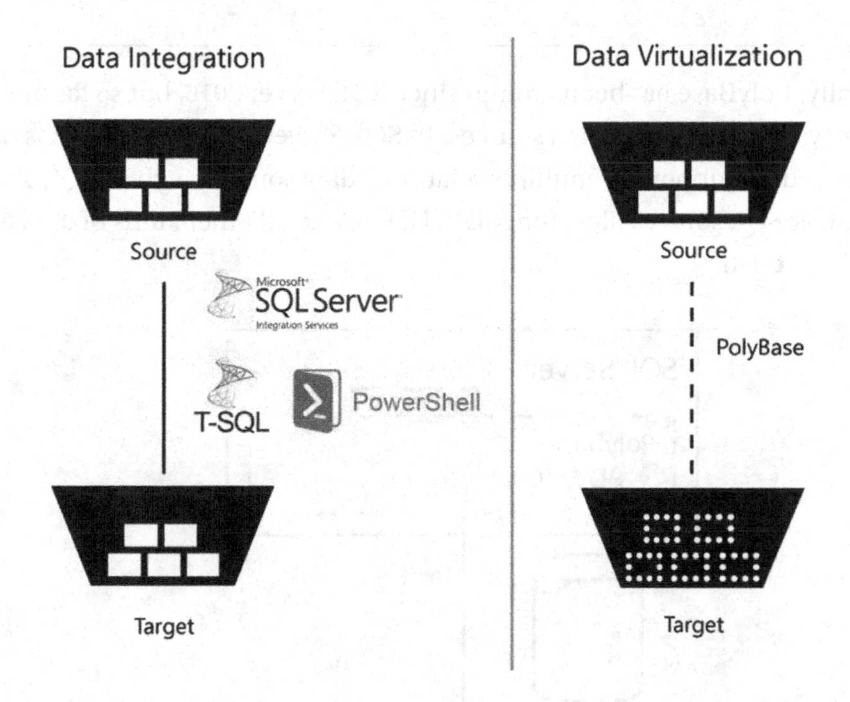


***Figure 1-2.*** *Data virtualization vs. data integration*

One obvious upside to data virtualization is that you get rid of redundant data as you don't copy it from the source but read it directly from there. Especially in cases where you only read a big flat file once to aggregate it, there may be little to no use to that duplicate and redundant data. Also, with PolyBase, your query is real time, whereas integrated data will always carry some lag.

On the other hand, you can't put indexes on an external table. Thus if you have data that you frequently query with different workloads than on the original source, which means that you require another indexing strategy, it might still make sense to integrate the data rather than virtualize it. That decision may also be driven by the question on whether you can accept the added workload to your source that would result from more frequent reporting queries and so on.

**Note** While data virtualization solves a couple of issues that come with data integration, it won't be able to replace data integration. This is NOT the end of SSIS or ETL 😉.

Technically, PolyBase has been around since SQL Server 2016, but so far only supported very limited types of data sources. In SQL Server 2019, PolyBase has been greatly enhanced by support for multiple relational data sources such as SQL Server or Oracle and NoSQL sources like MongoDB, HDFS, and all other kinds of data as we illustrate in Figure 1-3.

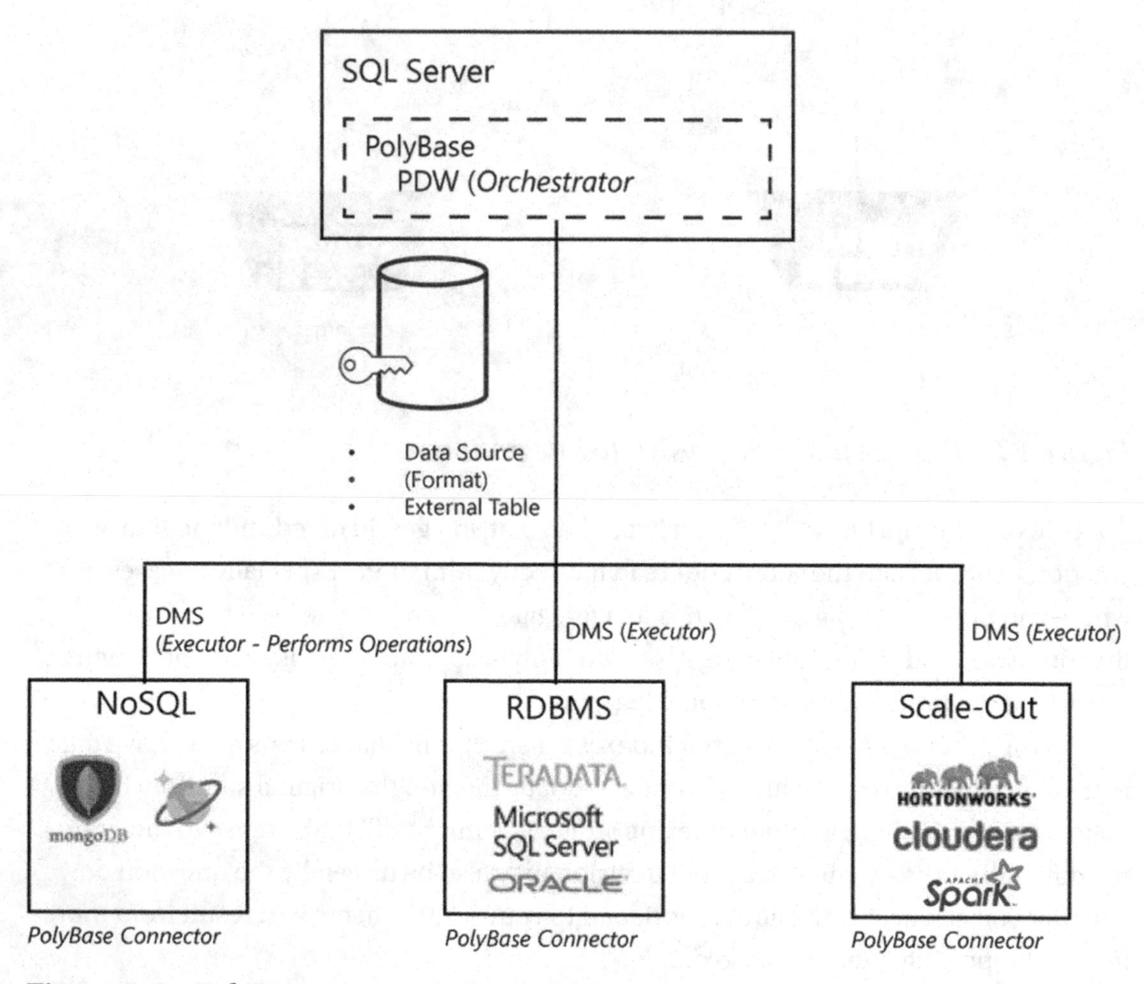


***Figure 1-3.*** *PolyBase sources and capabilities in SQL Server 2019*

Effectively, you can query a table in another database or even on a completely different machine as if it were a local table.

The use of PolyBase for virtualization may remind you of a linked server and there definitely are some similarities. One big difference is that the query toward a linked server tends to be longer and more involved than a PolyBase query. For example, here is a typical query against a remote table:

```
SELECT * FROM MyOtherServer.MyDatabase.DBO.MyTable
```

Using PolyBase, you would write the same query more simply, as if the table were in your local database. For example:

```
SELECT * FROM MyTable
```

PolyBase will know that the table is in a different database because you will have created a definition in PolyBase indicating where the table can be found.

An advantage of using PolyBase is that you can move *MyDatabase* to another server without having to rewrite your queries. Simply change your PolyBase data source definition to redirect to the new data source. You can do that easily, without harming or affecting your existing queries or views.

There are more differences between the use of linked servers and PolyBase. Table 1-1 describes some that you should be aware of.

***Table 1-1.*** *Comparison of linked servers and PolyBase*

| Linked Server | PolyBase |
| --- | --- |
| – Instance scoped | – Database scoped |
| – OLEDB providers | – ODBC drivers |
| – Read/write and pass-through statements | – Read-only operations |
| – Single-threaded | – Queries can be scaled out |
| – Separate configuration needed for each instance in Always On Availability Group | – No separate configuration needed for Always On Availability Group |

## Outsource Your Data

You may have heard of "Stretch Database,"[1] a feature introduced in SQL Server 2016, which allows you to offload parts of your data to Azure. The idea is to use the feature for "cold data" – meaning data that you don't access as frequently because it's either old (but still needed for some queries) or simply for business areas that require less attention.

The rationale behind cold data is that it should be cheaper to store that data in Azure than on premise. Unfortunately, the service may not be right for everyone as even its entry tier provides significant storage performance which obviously comes at a cost.

With PolyBase, you can now, for example, offload data to an Azure SQL Database and build your own very low-level outsourcing functionality.

## Reduce Data Redundancy and Development Time

Besides offloading data, the reason to virtualize it instead of integrating it is obviously the potentially tremendous reduction of data redundancy. As data virtualization keeps the data at its original source and the data is therefore not persisted at the destination, you basically cut your storage needs in half compared to a traditional ETL-based staging process.

---

**Note** Our "cut in half" assertion may not be super accurate as you may not have staged the full dataset anyway (reducing the savings) or you may have used different datatypes (potentially even increasing the savings even more).

---

Think of this: You want to track the number of page requests on your website per hour which is logging to text files. In a traditional environment, you would have written a SQL Server Integration Services (SSIS) package to load the text file into a table, then run a query on it to group the data, and then store or use its result. In this then new virtualization approach, you would still run the query to group the data but you'd run it right on your flat file, saving the time it would have taken to develop the SSIS package and also the storage for the staging table holding the log data which would otherwise have coexisted in the file as well as the staging table in SQL Server.

[1] https://azure.microsoft.com/en-us/pricing/details/sql-server-stretch-database/

## A Combined Data Platform Environment

One of the big use cases of SQL Server Big Data Clusters is the ability to create an environment that stores, manages, and analyzes data in different formats, types, and sizes. Most notably, you get the ability to store both relational data inside the SQL Server component and nonrelational data inside the HDFS storage subsystem. Using Big Data Clusters allows you to create a data lake environment that can answer all your data needs without a huge layer of complexity that comes with managing, updating, and configuring various parts that make up a data lake.

Big Data Clusters completely take care of the installation and management of your Big Data Cluster straight from the installation of the product. Since Big Data Clusters is being pushed as a stand-alone product with full support from Microsoft, this means Microsoft is going to handle updates for all the technologies that make up Big Data Clusters through service packs and updates.

So why would you be interested in a data lake? As it turns out, many organizations have a wide variety of data stored in different formats. In many situations, a large portion of data comes from the use of applications that store their data inside relational databases like SQL Server. By using a relational database, we can easily query the data inside of it and use it for all kinds of things like dashboards, KPIs, or even machine learning tasks to predict future sales, for instance.

A relational database must follow a number of rules, and one of the most important of those rules is that a relational database always stores data in a *schema-on-write* manner. This means that if you want to insert data into a relational database, you have to make sure the data complies to the structure of the table being written to. Figure 1-4 illustrates schema-on-write.

For instance, a table with the columns OrderID, OrderCustomer, and "OrderAmount" dictates that data you are inserting into that table will also need to contain those same columns. This means that when you want to write a new row in this table, you will have to define an OrderID, OrderCustomer, and OrderAmount for the insert to be successful. There is no room for adding additional columns on the fly, and in many cases, the data you are inserting needs to be the same datatype as specified in the table (for inside integers for numbers and strings for text).

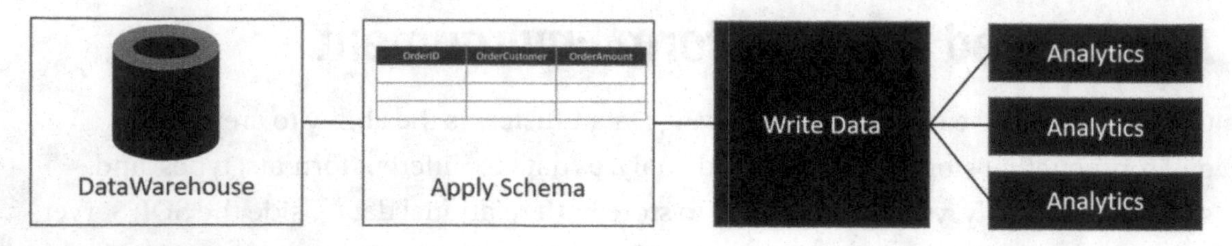


***Figure 1-4.*** *Scheme-on-write*

Now in many situations the schema-on-write approach is perfectly fine. You make sure all your data is formatted in the way the relational databases expect it to be, and you can store all your data inside of it. But what happens when you decide to add new datasets that do not necessarily have a fixed schema? Or, you want to process data that is very large (multiple terabytes) in terms of size? In those situations, it is frequently advised to look for another technology to store and process your data since a relational database has difficulties handling data with those characteristics.

Solutions like Hadoop and HDFS were created to solve some of the limitations around relational databases. Big Data platforms are able to process large volumes of data in a distributed manner by spreading the data across different machines (called nodes) that make up a cluster architecture. Using a technology like Hadoop, or as we will use in this book Spark, allows you to store and process data in any format. This means we can store huge CSV (comma-separated values) files, video files, Word documents, PDFs, or whatever we please without having to worry about complying to a predefined schema like we'd have to when storing data inside a relational database.

Apache's Spark technology makes sure our data is cut up into smaller blocks and stored on the filesystem of the nodes that make up a Spark cluster. We only have to worry about the schema when we are going to read in and process the data, something that is called *schema-on-read*. When we load in our CSV file to check its contents, we have to define what type of data it is and, in the case of a CSV file, what the columns are of the data. Specifying these details on read allows us a lot of flexibility when dealing with this data, since we can add or remove columns or transform datatypes without having to worry about a schema before we write the data back again. Because a technology like Spark has a distributed architecture, we can perform all these data manipulation and querying steps very quickly on large datasets, something we are explaining in more detail in Chapter 2.

What you see in the real world is that in many situations organizations have both relational databases and a Hadoop/Spark cluster to store and process their data. These solutions are implemented separately from each other and, in many cases, do not "talk" to each other. Is the data relational? Store it in the database! Is it nonrelational like CSV, IoT data, or other formats? Throw it on the Hadoop/Spark cluster! One reason why we are so excited over the release of SQL Server Big Data Clusters is that it combines both these solutions into a single product, a product that contains both the capabilities of a Spark cluster together with SQL Server. And while you still must choose whether you are going to store something directly in the SQL Server database or store it on the HDFS filesystem, you can always access it from both technologies! Want to combine relational data with a CSV file that is stored on HDFS? No problem, using data virtualization we described earlier in this chapter, you can read the contents from the CSV file from HDFS and merge it with your relational data using a T-SQL query producing a single result!

In this sense, SQL Server Big Data Clusters are made up from technologies that complement each other very well, allowing you to bridge the gap on how limited you are in processing data based on the manner in which it is stored. Big Data Clusters ultimately let you create a scalable and flexible data lake environment in which you can store and process data in any format, shape, or size, even allowing you to choose between processing the data using SQL Server or Spark, whichever you prefer for the tasks you want to perform.

The Big Data Cluster architecture will also be able to optimize performance in terms of data analytics. Having all data you require stored inside a single cluster, whether it is relational or not, means that you can access it immediately whenever you require it. You avoid data movement across different systems or networks, which is a huge advantage in a world where we are constantly trying to find solutions to analyze data faster and faster.

If you ask us what the ultimate advantage of SQL Server Big Data Clusters is, we firmly believe it is the ability to store, process, and analyze data in any shape, size, or type inside a single solution.

## Centralized AI Platform

As we described in the preceding section, SQL Server Big Data Clusters allow you to create a data lake environment that can handle all types and formats of data. Next to having huge advantages when processing, it naturally also has immense advantages when dealing with advanced analytics like machine learning. Since all your data is essentially stored in one place, you can perform tasks like machine learning model training on all the data that is available on the Big Data Cluster, instead of having to gather data from multiple systems across your organization.

By combining SQL Server and Spark, we also have multiple options available when working with machine learning. We can choose to train and score machine learning models through Spark directly by accessing data that is stored on the HDFS filesystem, or use the In-Database Machine Learning Services available to us through SQL Server. Both these options allow a wide variety in languages and libraries you, or your data science team, can use, for instance, R, Python, and Java for SQL Server Machine Learning Services, or PySpark and Scala when running your machine learning workload through the Spark cluster.

In terms of use cases, Big Data Clusters can handle just about any machine learning process, from handling real-time scoring to using GPUs in combination with TensorFlow to optimize the handling of CPU-intensive workloads or, for instance, perform image classification tasks.

CHAPTER 2

# Big Data Cluster Architecture

SQL Server Big Data Clusters are made up from a variety of technologies all working together to create a centralized, distributed data environment. In this chapter, we are going to look at the various technologies that make up Big Data Clusters through two different views.

First, we are evaluating the more-or-less physical architecture of Big Data Clusters. We are going to explore the use of containers, the Linux operating system, Spark, and the HDFS storage subsystem that make up the storage layer of Big Data Clusters.

In the second part of this chapter, we are going to look at the logical architecture which is made up of four different logical areas. These areas combine several technologies to provide a specific function, or role(s), inside the Big Data Cluster.

## Physical Big Data Cluster Infrastructure

The physical infrastructure of Big Data Clusters is made up from containers on which you deploy the major software components. These major components are SQL Server on Linux, Apache Spark, and the HDFS filesystem. The following is an introduction to these infrastructure elements, beginning with containers and moving through the others to provide you with the big picture of how the components fit together.

### Containers

A *container* is a kind of stand-alone package that contains everything you need to run an application in an isolated or sandbox environment. Containers are frequently compared to virtual machines (VMs) because of the virtualization layers that are present in both solutions. However, containers provide far more flexibility than virtual machines. A notable area of increased flexibility is the area of portability.




B. Weissman and E. van de Laar, *SQL Server Big Data Clusters*,
https://doi.org/10.1007/978-1-4842-5985-6_2

One of the main advantages of using containers is that they avoid the implementation of an operating system inside the container. Virtual machines require the installation of their own operating system inside each virtual machine, whereas with containers, the operating system of the host on which the containers are being run is used by each container (through isolated processes). Tools like Docker enable multiple operating systems on a single host machine by running a virtual machine that becomes the host for your containers, allowing you to run a Linux container on Windows, for example.

You can immediately see an advantage here: when running several virtual machines, you also have an additional workload of maintaining the operating system on each virtual machine with patches, configuring it, and making sure everything is running the way it is supposed to be. With containers, you do not have those additional levels of management. Instead, you maintain one copy of the operating system that is shared among all containers.

Another advantage for containers over virtual machines is that containers can be defined as a form of "infrastructure-as-code." This means you can script out the entire creation of a container inside a build file or image. This means that when you deploy multiple containers with the same image or build file, they are 100% identical. Ensuring 100% identical deployment is something that can be very challenging when using virtual machines, but is easily done using containers.

Figure 2-1 shows some differences between containers and virtual machines around resource allocation and isolation. You can see how containers reduce the need for multiple guest operating systems.

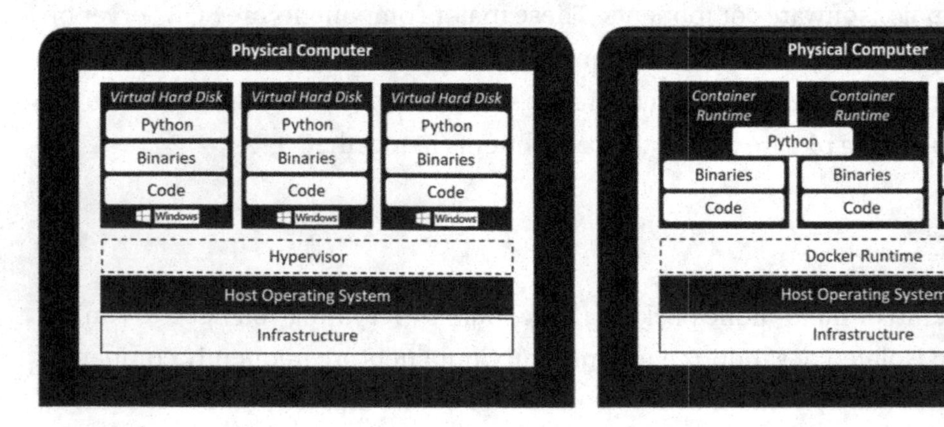


***Figure 2-1.*** *Virtual machine vs. containers*

A final advantage of containers we would like to mention (there are many more to name, however, that would go beyond the scope of this book) is that containers can be deployed as "stateless" applications. Essentially this means that containers won't change, and they do not store data inside themselves.

Consider, for instance, a situation in which you have a number of application services deployed using containers. In this situation, each of the containers would run the application in the exact same manner and state as the other containers in your infrastructure. When one container crashes, it is easy to deploy a new container with the same build file filling in the role of the crashed container, since no data inside the containers is stored or changed for the time they are running.

The storage of your application data could be handled by other roles in your infrastructure, for instance, a SQL Server that holds the data that is being processed by your application containers, or, as a different example, a file share that stores the data that is being used by the applications inside your containers. Also, when you have a new software build available for your application servers, you can easily create a new container image or build file, map that image or build file to your application containers, and switch between build versions practically on the fly.

SQL Server Big Data Clusters are deployed using containers to create a scalable, consistent, and elastic environment for all the various roles and functions that are available in Big Data Clusters. Microsoft has chosen to deploy all the containers using Kubernetes. Kubernetes is an additional layer in the container infrastructure that acts like an orchestrator. By using Kubernetes (or K8s as it is often called), you get several advantages when dealing with containers. For instance, Kubernetes can automatically deploy new containers whenever it is required from a performance perspective, or deploy new containers whenever others fail.

Because Big Data Clusters are built on top of Kubernetes, you have some flexibility in where you deploy Big Data Clusters. Azure has the ability to use a managed Kubernetes Service (AKS) where you can also choose to deploy Big Data Clusters if you so want to. Other, on-premises options are Docker or Minikube as container orchestrators. We will take a more in-depth look at the deployment of Big Data Clusters inside AKS, Docker, or Minikube in Chapter 3.

Using Kubernetes also introduces a couple of specific terms that we will be using throughout this book. We've already discussed the idea and definition of containers. However, Kubernetes (and also Big Data Clusters) also frequently uses another term called "pods." Kubernetes does not run containers directly; instead it wraps a container in a structure called a pod. A pod combines one or multiple containers, storage resources, networking configurations, and a specific configuration governing how the container should run inside the pod.

Figure 2-2 shows a simple representation of the node – pods – container architecture inside Kubernetes.

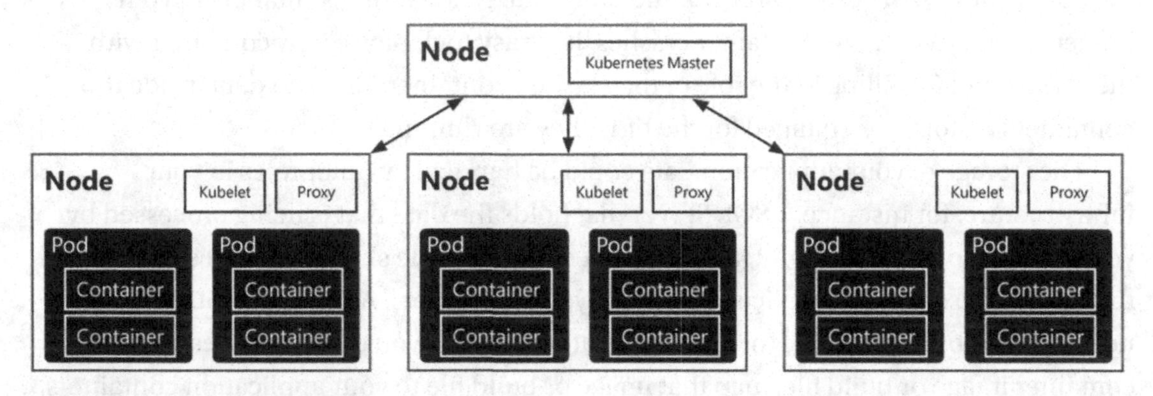


***Figure 2-2.*** *Representation of containers, pods, and nodes in Kubernetes*

Generally, pods are used in two manners: a single container per pod or multiple containers inside a single pod. The latter is used when you have multiple containers that need to work together in one way or the other – for instance, when distributing a load across various containers. Pods are also the resource managed to allocate more system resources to containers. For example, to increase the available memory for your containers, a change in the pod's configuration will result in access to the added memory for all containers inside the pod. On that note, you are mostly managing and scaling pods instead of containers inside a Kubernetes cluster.

Pods run on Kubernetes nodes. A node is the smallest unit of computing hardware inside the Kubernetes cluster. Most of the time, a node is a single physical or virtual machine on which the Kubernetes cluster software is installed, but in theory every machine/device with a CPU and memory can be a Kubernetes node. Because these machines only function as hosts of Kubernetes pods, they can easily be replaced, added, or removed from the Kubernetes architecture, making the underlying physical (or virtual) machine infrastructure very flexible.

## SQL Server on Linux

In March 2016, Microsoft announced that the next edition of SQL Server, which turned out to be SQL Server 2017, would be available not only on Windows operating systems but on Linux as well – something that seemed impossible for as long as Microsoft has been building software suddenly became a reality and, needless to say, the entire IT world freaked out.

In hindsight, Microsoft had perfect timing in announcing the strategic decision to make one of its flagship products available on Linux. The incredible adaptation of new technologies concerning containers, which we discussed in the previous section, was mostly based on Linux distributions. We believe that without the capability's containers, and thus the Linux operating system those containers provide, there would never have been a SQL Server Big Data Cluster product.

Thankfully Microsoft pushed through on their adoption of Linux, and with the latest SQL Server 2019 release, many of the issues that plagued the SQL Server 2017 release on Linux are now resolved and many capabilities that were possible on the Windows version have been brought to Linux as well.

So how did Microsoft manage to run an application designed for the Windows operating system on Linux? Did they rewrite all the code inside SQL Server to make it run on Linux? As it turns out, things are far more complicated than a rewrite of the code base to make it Linux compatible.

To make SQL Server run on Linux, Microsoft introduced a concept called a Platform Abstraction Layer (or PAL for short). The idea of a PAL is to separate the code needed to run, in this case, SQL Server from the code needed to interact with the operating system. Because SQL Server has never run on anything other than Windows, SQL Server is full of operating system references inside its code. This would mean that getting SQL Server to run on Linux would end up taking enormous amounts of time because of all the operating system dependencies.

The SQL Server team looked for different approaches to resolve this issue of operating system dependencies and found their answer in a Microsoft research project called Drawbridge. The definition of Drawbridge can be found on its project page at `www.microsoft.com/en-us/research/project/drawbridge/` and states:

> *Drawbridge is a research prototype of a new form of virtualization for application sandboxing. Drawbridge combines two core technologies: First, a picoprocess, which is a process-based isolation container with a minimal kernel* API *surface. Second, a library OS, which is a version of Windows enlightened to run efficiently within a picoprocess.*

The main part that attracted the SQL Server team to the Drawbridge project was the library OS technology. This new technology could handle a very wide variety of Windows operating system calls and translate them to the operating system of the host, which in this case is Linux.

Now, the SQL Server team did not adapt the Drawbridge technology one-on-one as there were some challenges involved with the research project. One of the challenges was that the research project was officially completed which means that there was no support on the project. Another challenge was a large overlap of technologies inside the SQL Server OS (SOS) and Drawbridge. Both solutions, for example, have their own functionalities to handle memory management and threading/scheduling.

What eventually was decided was to merge the SQL Server OS and Drawbridge into a new platform layer called the SQLPAL (SQL Platform Abstraction Layer). Using SQLPAL, the SQL Server team can develop code as they have always done and leave the translation of operating system calls to the SQLPAL. Figure 2-3 shows the interaction between the various layers while running SQL Server on Linux.

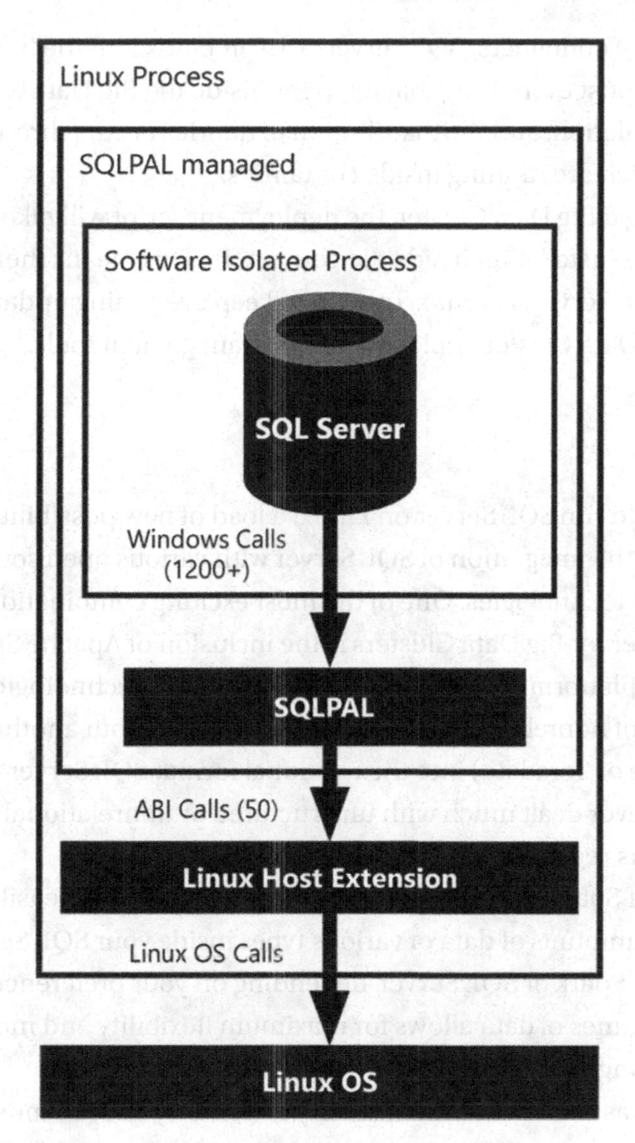


***Figure 2-3.** Interaction between the various layers of SQL Server on Linux*

There is a lot more information available on various Microsoft blogs that cover more of the functionality and the design choices of the SQLPAL. If you want to know more about the SQLPAL, or how it came to life, we would recommend the article "SQL Server on Linux: How? Introduction" available at `https://cloudblogs.microsoft.com/sqlserver/2016/12/16/sql-server-on-linux-how-introduction/`.

Next to the use of containers, SQL Server 2019 on Linux is at the heart of the Big Data Cluster product. Almost everything that happens inside the Big Data Cluster in terms of data access, manipulation, and the distribution of queries occurs through SQL Server on Linux instances which are running inside containers.

When deploying a Big Data Cluster, the deployment script will take care of the full SQL Server on Linux installation inside the containers. This means there is no need to manually install SQL Server on Linux, or even to keep everything updated. All of this is handled by the Big Data Cluster deployment and management tools.

## Spark

With the capability to run SQL Server on Linux, a load of new possibilities became available regarding the integration of SQL Server with various open source and Linux-based products and technologies. One of the most exciting combinations that became a reality inside SQL Server Big Data Clusters is the inclusion of Apache Spark.

SQL Server is a platform for relational databases. While technologies like PolyBase enable the reading of nonrelational data (or relational data from another relational platform like Oracle or Teradata) into the relational format SQL Server requires, in its heart SQL Server never dealt much with unstructured or nonrelational data. Spark is a game changer in this regard.

The inclusion of Spark inside the product means you can now easily process and analyze enormous amounts of data of various types inside your SQL Server Big Data Cluster using either Spark or SQL Server, depending on your preferences. This ability to process large volumes of data allows for maximum flexibility and makes parallel and distributed processing of datasets a reality.

Apache Spark was created at the University of Berkeley in 2009 mostly as an answer to the limitations of a technology called MapReduce. The MapReduce programming model was developed by Google and was the underlying technology used to index all the web pages on the Internet (and might be familiar to you in the Hadoop MapReduce form). MapReduce is best described as a framework for the parallel processing of huge datasets using a (large) number of computers known as nodes. This parallel processing across multiple nodes is important since datasets reached such sizes that they could no longer efficiently be processed by single machines. By spreading the work, and data, across multiple machines, parallelism could be achieved which results in the faster processing of those big datasets.

Running a query on a MapReduce framework usually results in going through four steps of execution:

1. Input splits

   The input to a MapReduce job is split into logical distribution of the data stored in file blocks. The MapReduce job calculates which records fit inside a logical block, or "splits," and decides on the number of mappers that are required to process the job.

2. Mapping

   During mapping our query is being performed on each of the "splits" separately and produces the output for the specific query on the specific split. The output is always a form of key/value pairs that are returned by the mapping process.

3. Shuffling

   The shuffling process is, simply said, the process of sorting and consolidating the data that was returned by the mapping process.

4. Reducing

   The final step, reducing, aggregates the results returned by the shuffling process and returns a single output.

The best way to explain the MapReduce process is by looking at a visual example. Figure 2-4 shows an example of a MapReduce task that calculates word occurrences inside a dataset. To keep things simple and visually easy to display, we use a simple, short sentence that acts as a dataset: "SQL Server is no longer just SQL but it is much more."

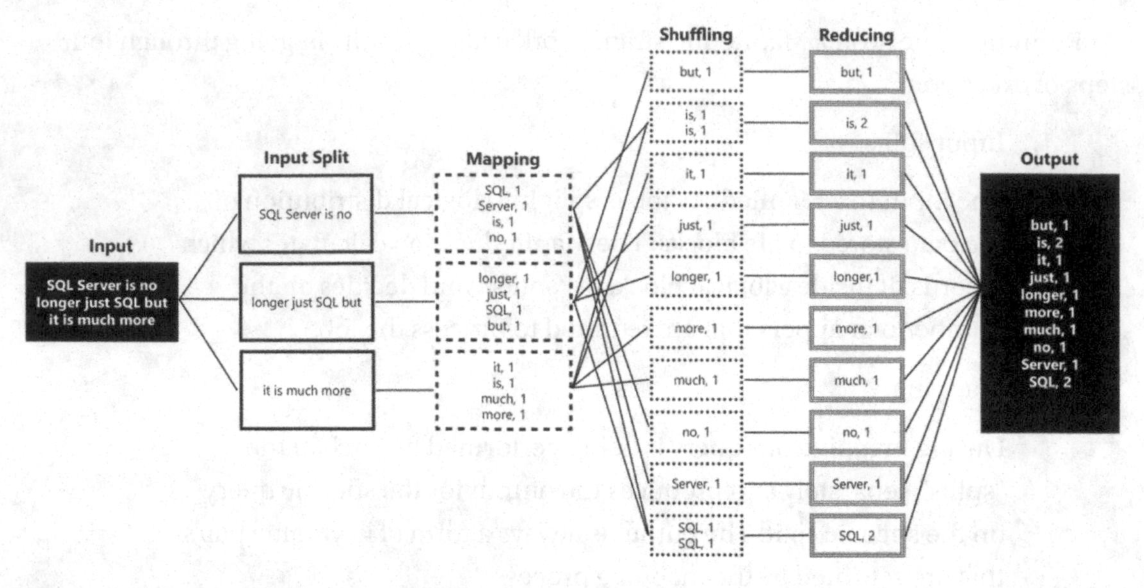


***Figure 2-4.** MapReduce example job*

What happens in the example is that the dataset that contains the input for our job (the sentence "SQL Server is no longer just SQL but is also much more") is split up into three different splits. These splits are processed in the mapping phase, resulting in the word counts for each split. The results are sent to the shuffling step which places all the results in order. Finally, the reduce step calculates the total occurrences for each individual word and returns it as the final output.

As you can see from the (simple) example in Figure 2-4, MapReduce is very efficient in distributing and coordinating the processing of data across a parallel environment. However, the MapReduce framework also had a number of drawbacks, the most notable being the difficulty of writing large programs that require multiple passes over the data (for instance, machine learning algorithms). For each pass over a dataset, a separate MapReduce job had to be written, each one loading the data it required from scratch again. Because of this, and the way MapReduce accesses data, processing data inside the MapReduce framework can be rather slow.

Spark was created to address these problems and make big data analytics more flexible and better performing. It does so by implementing in-memory technologies that allow sharing of data between processing steps and by allowing ad hoc data processing instead of having to write complex MapReduce jobs to process data. Also, Spark supports a wide variety of libraries that can enhance or expand the capabilities of Spark, like processing streaming data or performing machine learning tasks, and even query data through the SQL language.

Spark looks and acts a lot like the MapReduce framework in that Spark is also a coordinator, and manager, of tasks that process data. Just like MapReduce, Spark uses workers to perform the actual processing of data. These workers get told what to do through a so-called Spark application which is defined as a driver process. The driver process is essentially the heart of a Spark application, and it keeps track of the state of your Spark application, responds to input or output, and schedules and distributes work to the workers. One advantage of the driver process is that it can be "driven" from different programming languages, like Python or R, through language APIs. Spark handles the translation of the commands in the various languages to Spark code that gets processed on the workers.

Figure 2-5 shows an overview of the *logical* Spark architecture.

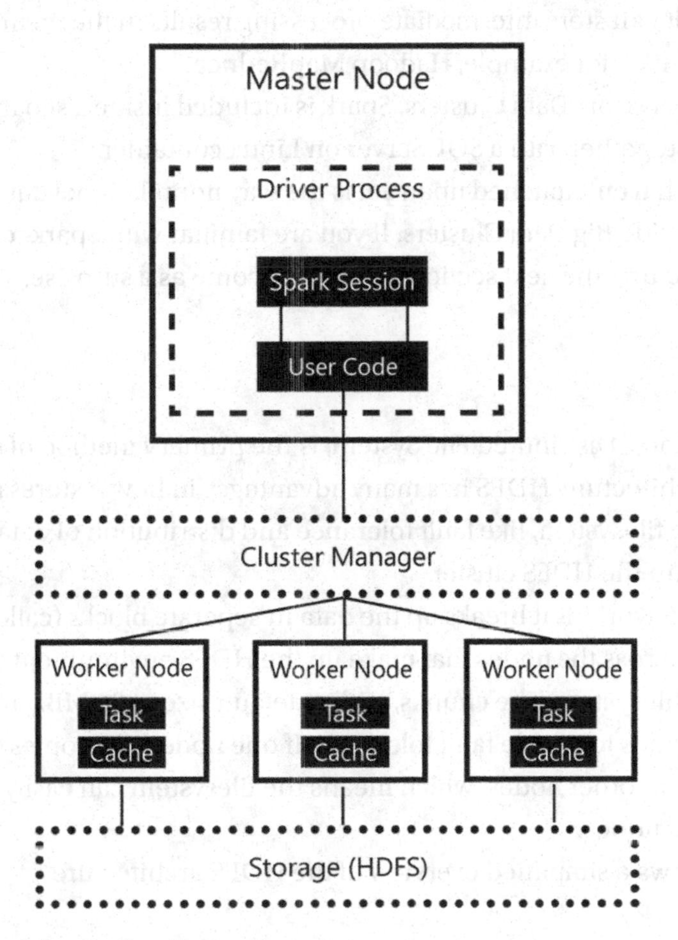


***Figure 2-5.** Spark logical architecture*

There is a reason why we specifically mentioned the word "logical" in connection with Spark's architecture. Even though Figure 2-5 implies that worker nodes are separate machines that are part of a cluster, it is in fact possible in Spark to run as many worker nodes on a machine as you please. As a matter of fact, both the driver process and worker nodes can be run on a single machine in local mode for testing and development tasks.

Figure 2-5 also shows how a Spark application coordinates work across the cluster. The code you write as a user is translated by the driver process to a language your worker nodes understand; it distributes the work to the various worker nodes which handle the data processing. In the illustration, we specifically highlighted the cache inside the worker node. The cache is one part of why Spark is so efficient in performing data processing since it can store intermediate processing results in the memory of the node, instead of on disk like, for example, Hadoop MapReduce.

Inside SQL Server Big Data Clusters, Spark is included inside a separate container that shares a pod together with a SQL Server on Linux container.

One thing we haven't touched upon yet is the way nonrelational data outside SQL Server is stored inside Big Data Clusters. If you are familiar with Spark- or Hadoop-based big data infrastructure, the next section should not come as a surprise.

## HDFS

HDFS, or the Hadoop Distributed File System, is the primary method of storing data inside a Spark architecture. HDFS has many advantages in how it stores and processes data stored on the filesystem, like fault tolerance and distribution of data across multiple nodes that make up the HDFS cluster.

The way HDFS works is it breaks up the data in separate blocks (called chunks) and distributes them across the nodes that make up the HDFS environment when data is stored inside the filesystem. The chunks, with a default size of 64 MB, are then replicated across multiple nodes to enable fault tolerance. If one node fails, copies of data chunks are also available on other nodes, which means the filesystem can easily recover from data loss on single nodes.

Figure 2-6 shows a simplified overview of the HDFS architecture.

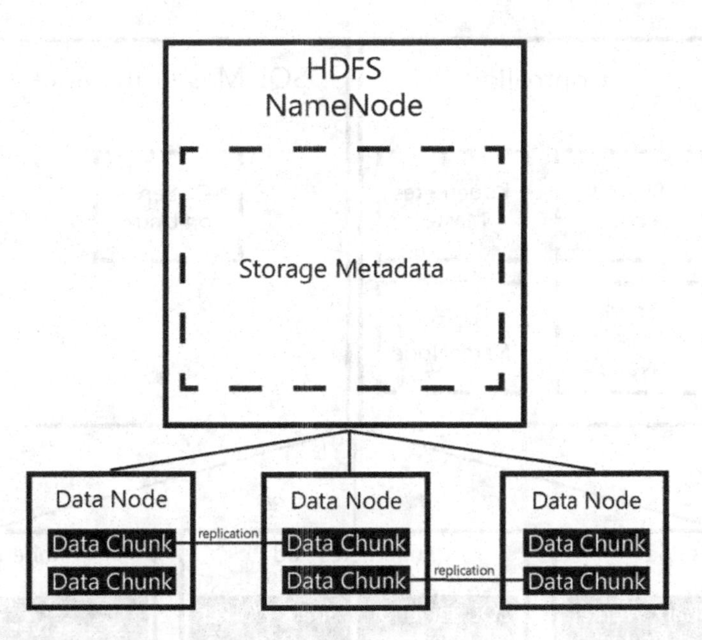


***Figure 2-6.*** *HDFS architecture*

In many aspects, HDFS mirrors the architecture of Hadoop and, in that sense, of Spark as we have shown in the previous section. Because of the distributed nature of data stored inside the filesystem, it is possible, and in fact expected, that data is distributed across the nodes that also handle the data processing inside the Spark architecture. This distribution of data brings a tremendous advantage in performance; since data resides on the same node that is responsible for the processing of that data, it is unnecessary to move data across a storage architecture. With the added benefit of data caching inside of the Spark worker nodes, data can be processed very efficiently indeed.

One thing that requires pointing out is that unlike with Hadoop, Spark is not necessarily restricted to data residing in HDFS. Spark can access data that is stored in a variety of sources through APIs and native support. Examples include various cloud storage platforms like Azure Blob Storage or relational sources like SQL Server.

## Tying the Physical Infrastructure Parts Together

Now it's time to look at the big picture. Figure 2-7 shows a complete overview of how the technologies discussed in the previous sections work together inside SQL Server Big Data Clusters.

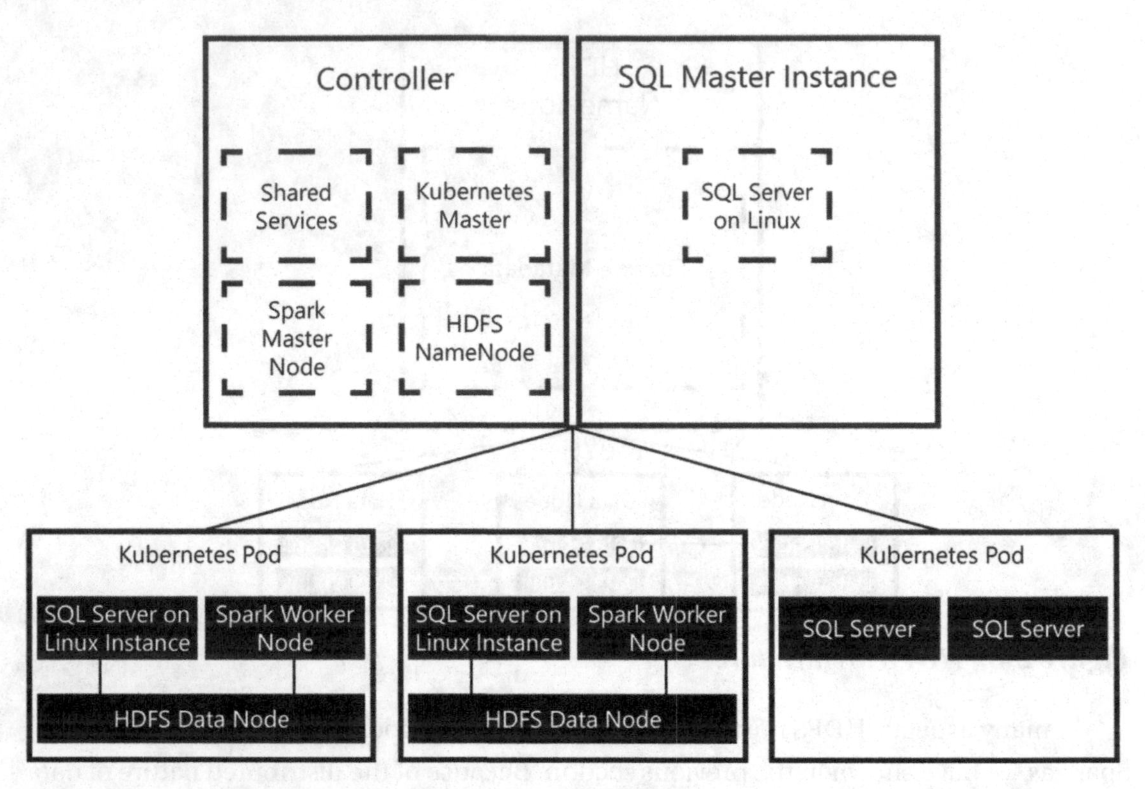


***Figure 2-7.*** *SQL Server Big Data Cluster architecture with Spark, HDFS, and Kubernetes*

As you can see from Figure 2-7, SQL Server Big Data Clusters combine a number of different roles inside the containers that are deployed by Kubernetes. Placing both SQL Server and Spark together in a container with the HDFS Data Node allows both products to access data that is stored inside HDFS in a performance-optimized manner. In SQL Server this data access will occur through PolyBase, while Spark can natively query data residing in HDFS.

The architecture in Figure 2-7 also gives us two distinct different paths in how we can process and query our data. We can decide on storing data inside a relational format using the SQL Server instances available to us, or we can use the HDFS filesystem and store (nonrelational) data in it. When the data is stored in HDFS, we can access and process that data in whichever manner we prefer. If your users are more familiar with writing T-SQL queries to retrieve data, you can use PolyBase to bring the HDFS-stored data inside SQL Server using an external table. On the other hand, if users prefer to use Spark, they can write Spark applications that access the data directly from HDFS. Then if needed, users can invoke a Spark API to combine relational data stored in SQL Server with the nonrelational data stored in HDFS.

# Logical Big Data Cluster Architecture

As mentioned in the introduction of this chapter, Big Data Clusters can be divided into four logical areas. Consider these areas as a collection of various infrastructure and management parts that perform a specific function inside the cluster. Each of the areas in turn has one or more roles it performs. For instance, inside the Data Pool area, there are two roles: the Storage Pool and the SQL Data Pool.

Figure 2-8 shows an overview of the four logical areas and the various roles that are part of each area.

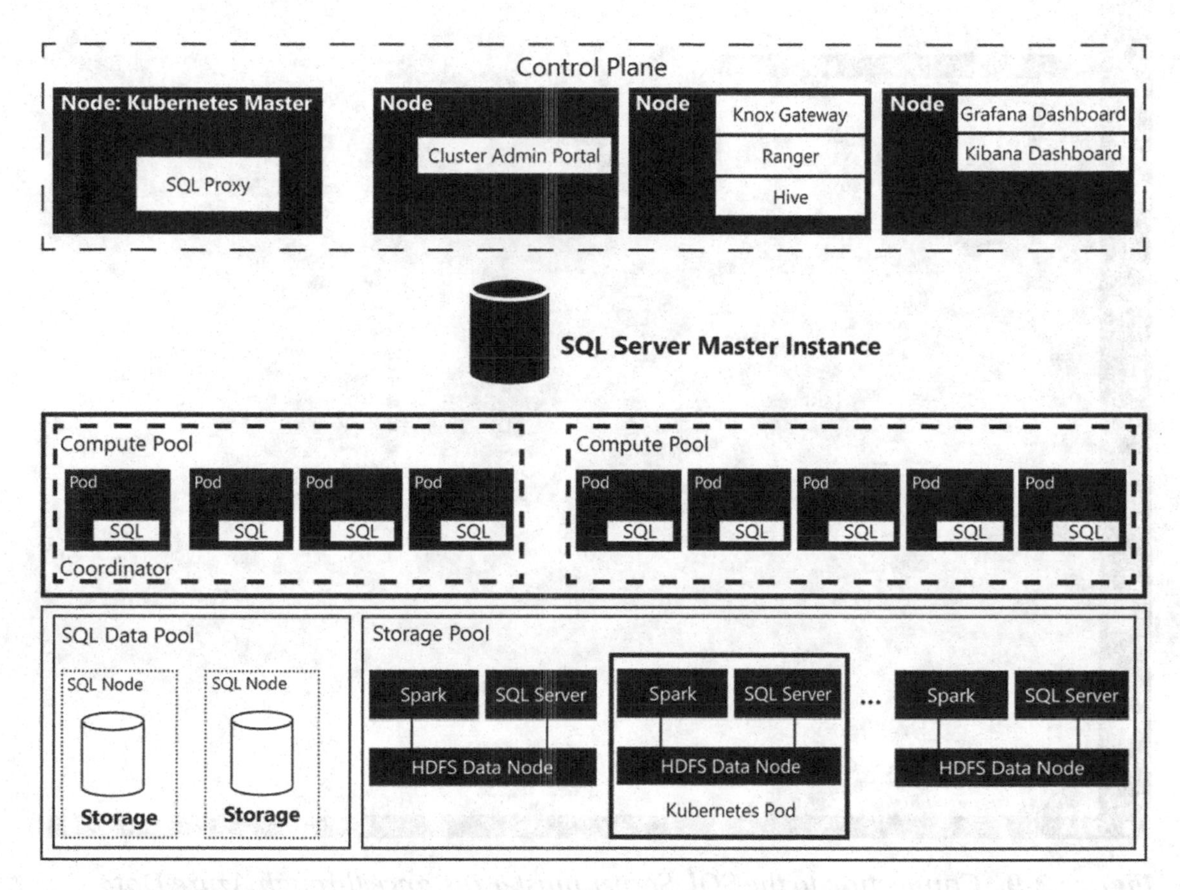


***Figure 2-8.** Big Data Cluster architecture*

You can immediately infer the four logical areas: the Control area (which internally is named the Control Plane) and the Compute, Data, and App areas. In the following sections, we are going to dive into each of these logical areas individually and describe their function and what roles are being performed in it. Before we start taking a closer look at the Control Plane, you might have noticed there is an additional role displayed in Figure 2-8, the SQL Server Master Instance.

The SQL Server master instance is a SQL Server on Linux deployment inside a Kubernetes node. The SQL Server master instance acts like an entry point toward your Big Data Cluster and provides the external endpoint to connect to through Azure Data Studio (ADS) (see Figure 2-9) or from other tools like SQL Server Management Studio.

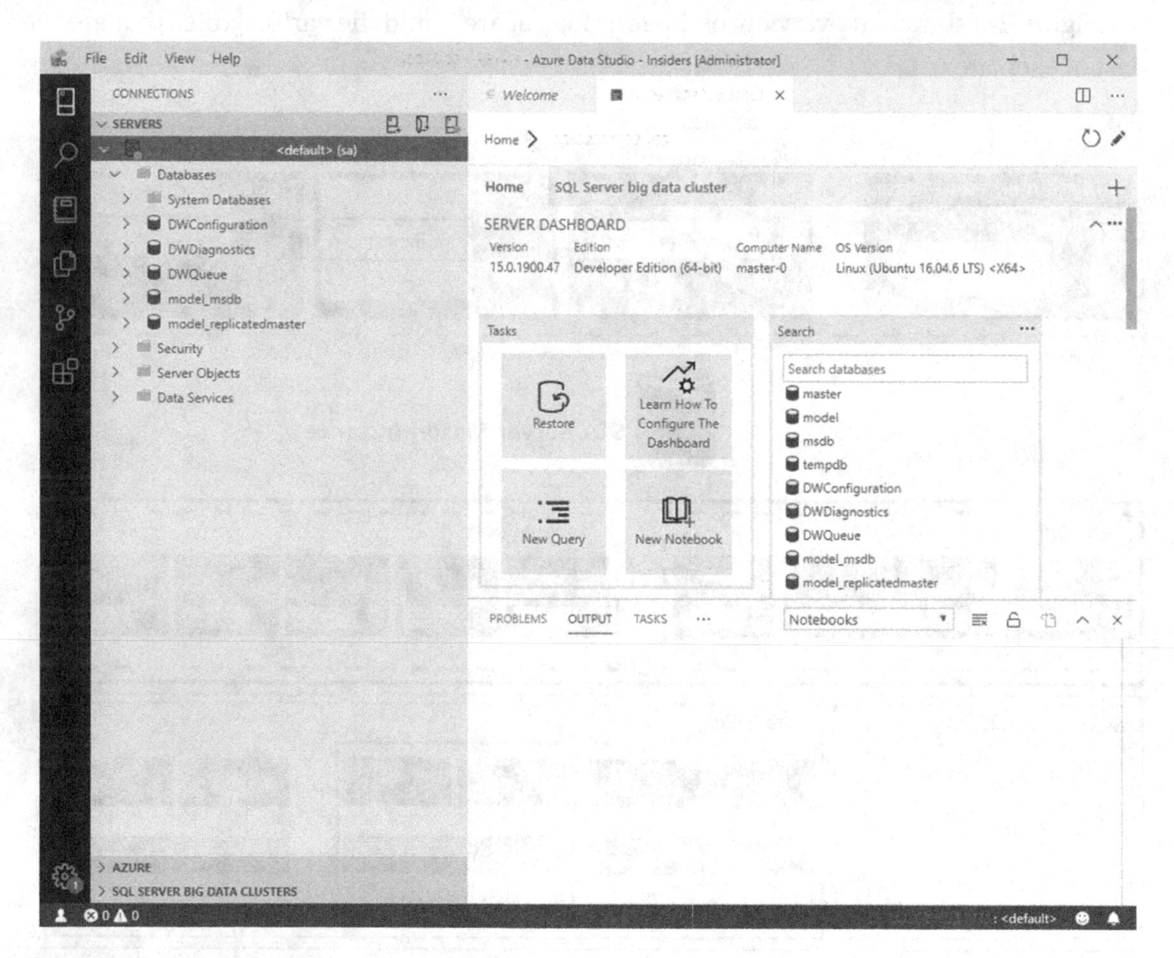


***Figure 2-9.** Connection to the SQL Server master instance through Azure Data Studio*

In many ways the SQL Server master instance acts like a normal SQL Server instance. You can access it and browse through the instance using Azure Data Studio and query the system and user databases that are stored inside of it. One of the big changes compared to a traditional SQL Server instance is that the SQL Server master instance will distribute your queries across all SQL Server nodes inside the Compute Pool(s) and access data that is stored, through PolyBase, on HDFS inside the Data Plane.

By default, the SQL Server master instance also has Machine Learning Services enabled. This allows you to run in-database analytics using R, Python, or Java straight from your queries. Using the data virtualization options provided in SQL Server Big Data Cluster, Machine Learning Services can also access nonrelational data that is stored inside the HDFS filesystem. This means that your data analysists or scientists can choose to use either Spark or SQL Server Machine Learning Services to analyze, or operationalize, the data that is stored in the Big Data Cluster. We are going to explore these options in a more detailed manner in Chapter 7.

# Control Plane

The Control Plane shown in Figure 2-10 is your entry into the Big Data Cluster management environment. It provides various management and log tools like Grafana and is the centralized place where you perform all your Big Data Cluster management. Also, security inside the Big Data Cluster is managed and controlled through the Control Plane.

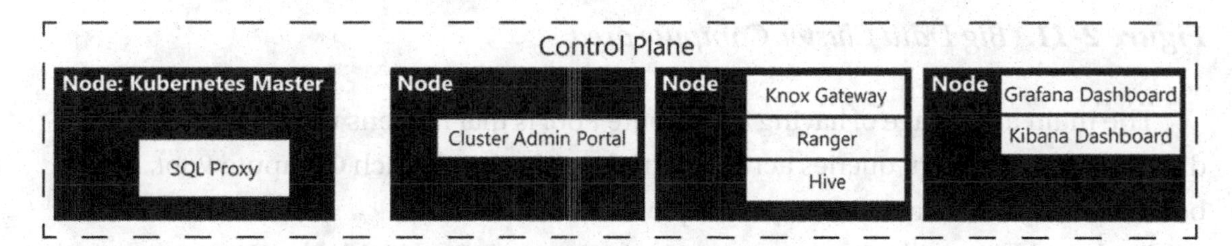


***Figure 2-10.*** *Big Data Cluster Control Plane*

In terms of managing Big Data Clusters, we are going to discuss the various management tools we can use to manage Big Data Clusters in Chapter 3.

Next to providing a centralized location where you can perform all your Big Data Cluster management tasks, the Control Plane also plays a very important part in the coordination of tasks to the underlying Compute and Data areas. The access to the Control Plane is available through the controller endpoint.

The controller endpoint is used for the Big Data Cluster management in terms of deployment and configuration of the cluster. The endpoint is accessed through REST APIs, and some services inside the Big Data Cluster, as well as the command-line tool we use to deploy and configure our Big Data Cluster, access those APIs.

You are going to get very familiar with the controller endpoint in the next chapter, in which we will deploy and configure a Big Data Cluster using **azdata**.

## Compute Area

The Compute area (see Figure 2-11) is made up from one or more Compute Pools. A Compute Pool is a collection Kubernetes Pods which contain SQL Server on Linux. Using a Compute Pool, you can access various data sources through PolyBase in a distributed manner. For instance, a Compute Pool can access data stored inside HDFS on the Big Data Cluster itself or access data through any of the PolyBase connectors like Oracle or MongoDB.

***Figure 2-11.*** *Big Data Cluster Compute area*

The main advantage of having a Compute Pool is that it opens up options to distribute, or scale out, queries across multiple nodes inside each Compute Pool, boosting the performance of PolyBase queries.

By default, you will have access to a single Compute Pool inside the Compute logical area. You can, however, add multiple Compute Pools in situations where, for instance, you want to dedicate resources to access a specific data source. All management and configuration of each Kubernetes Pod inside the Compute Pool is handled by the SQL Server Master Instance.

## Data Area

The Data area (Figure 2-12) is used to persist and cache data inside your Big Data Cluster, and it is split into two different roles, the Storage Pool and the SQL Data Pool, which both have different functionalities inside the Big Data Cluster.

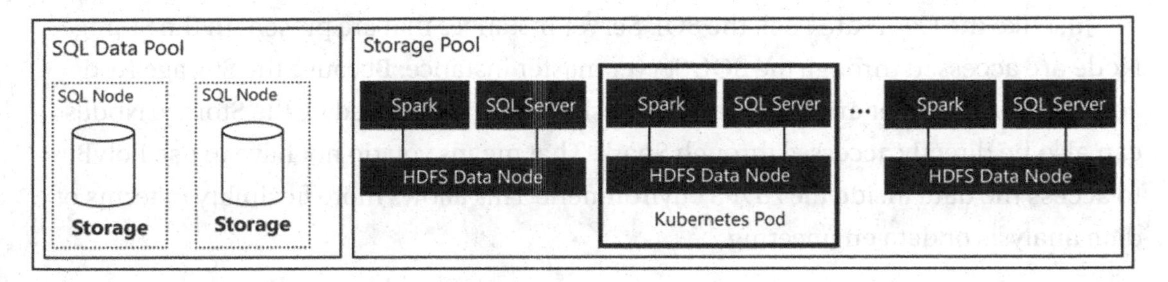


***Figure 2-12.** Data Plane architecture*

## Storage Pool

The Storage Pool consists of Kubernetes Pods that combine Spark, SQL Server on Linux, and a HDFS Data Node. Figure 2-13 illustrates the pod contents.

The HDFS Data Nodes are combined into a single HDFS cluster that is present inside your Big Data Cluster. The main function of the Storage Pool is to provide a HDFS storage cluster to store data on what is ingested through, for instance, Spark. By creating a HDFS cluster, you basically have access to a data lake inside the Big Data Cluster where you can store a wide variety of nonrelational data, like Parquet or CSV files.

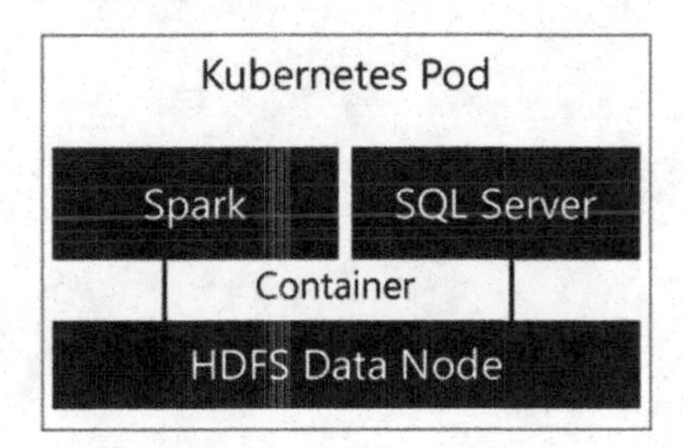


***Figure 2-13.** Storage Node inside the Storage Pool*

The HDFS cluster automatically arranges data persistence since the data you import into the Storage Pool is automatically spread across all the Storage Nodes inside the Storage Pool. This spreading of data across nodes also allows you to quickly analyze large volumes of data, since the load of analysis is spread across the multiple nodes. One advantage of this architecture is that you can either use the local storage present in the Data Node or add your own persistent storage subsystem to the nodes.

Just like the Compute Pool, the SQL Server instances that are present in the Storage Node are accessed through the SQL Server master instance. Because the Storage Node combines SQL Server and Spark, all data residing on or managed by the Storage Nodes can also be directly accessed through Spark. That means you do not have to use PolyBase to access the data inside the HDFS environment. This allows more flexibility in terms of data analysis or data engineering.

## SQL Data Pool

Another area of the Data Plane is the SQL Data Pool. This collection of pods is different compared to the Storage Pool in that it doesn't combine Spark or HDFS together into the node. Instead, the SQL Data Pool consists of one, or multiple, SQL Server on Linux instances. These instances are termed as shards, and you can see them illustrated in Figure 2-14.

The main role of the SQL Data Pool is to optimize access to external sources using PolyBase. The SQL Data Pool can than partition and cache data from those external sources inside the SQL Server instances and ultimately provide parallel and distributed queries against the external data sources. To provide this parallel and distributed functionality, datasets inside the SQL Data Pool are divided into shards across the nodes inside the SQL Data Pool.

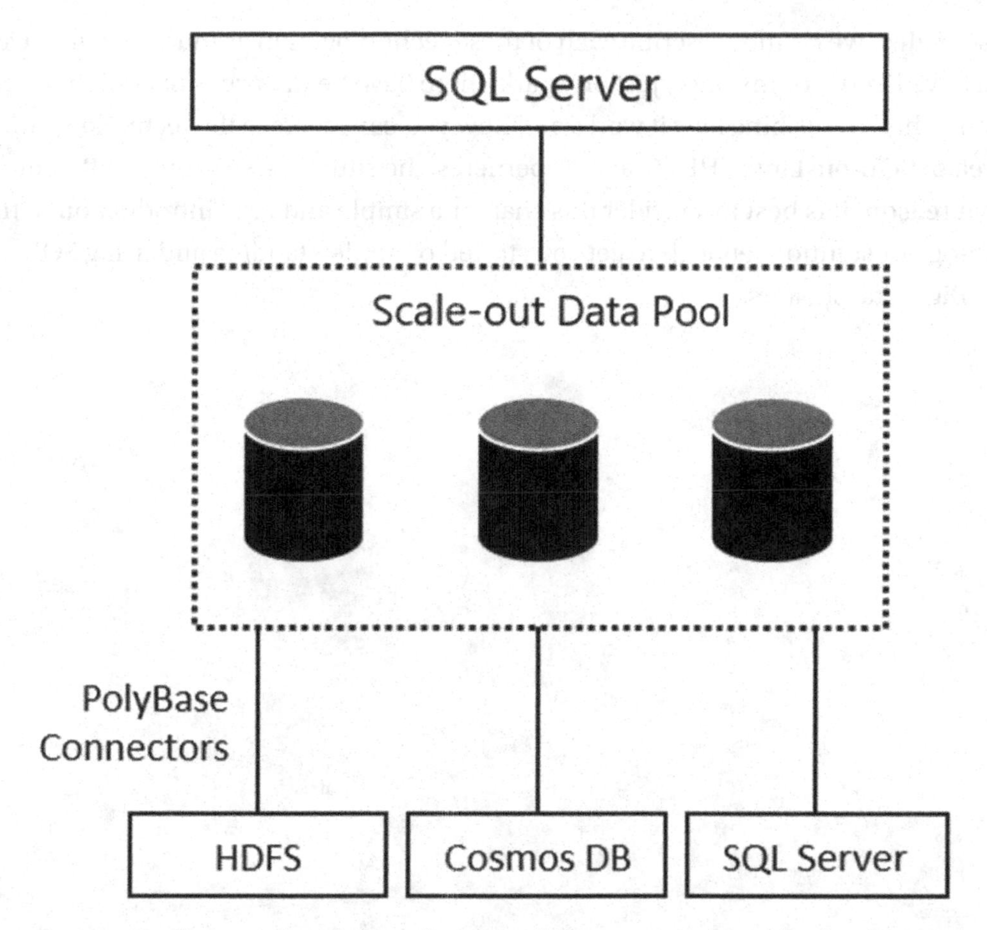


***Figure 2-14.*** *Scaling and caching of external data sources inside the SQL Data Pool*

# Summary

In this chapter, we've looked at the SQL Server Big Data Cluster architecture in two manners: physical and logical. In the physical architecture, we focused on the technologies that make up the Big Data Cluster like containers, SQL-on-Linux and Spark. In the logical architecture, we discussed the different logical areas inside Big Data Clusters that each perform a specific role or task inside the cluster.

For each of the technologies used in Big Data Clusters, we gave a brief introduction in its origins as well as what part the technology plays inside Big Data Clusters. Because of the wide variety of technologies and solutions used in SQL Server Big Data Clusters, we tried to be as thorough as possible in describing the various technologies; however, we also realize we cannot describe each of these technologies in as much detail as we would have liked. For instance, just on Spark, there have been dozens of books written and published describing how it works and how you can leverage the technology. In the area of SQL-on-Linux, HDFS, and Kubernetes, the situation isn't much different. For that reason, it is best to consider this chapter a simple and brief introduction to the technology or solution, enough to get you started on understanding and using SQL Server Big Data Clusters.

# CHAPTER 3

# Deployment of Big Data Clusters

Now it is time to install your very own SQL Server 2019 Big Data Cluster! We will be handling three different scenarios in detail and we will be using a fresh machine for each of those scenarios:

- Stand-alone PolyBase installation on Windows
- Big Data Cluster using kubeadm on Linux
- Big Data Cluster using Azure Kubernetes Service (AKS)

It is perfectly fine to run all options from the same box. But as it is likely that you will not be using all of them, we figured it would make sense to start fresh each time.

We will be covering the installation using the Microsoft Windows operating system. The goal of this guide is to get your Big Data Cluster up and running as quick as possible, so we will configure some options that may not be best practice (like leaving all the service accounts and directories on default). Feel free to modify those as it may fit your needs.

If you opt for the AKS installation, you will need an active Azure subscription. If you do not already have an Azure subscription, you can create one for free which includes credits which you can spend free of charge.

## A Little Helper: Chocolatey

Before we get started, we'd like to point your attention to Chocolatey – or choco. In case you haven't heard about it, choco is a free package manager for Windows which will allow us to install many of our prerequisites with a single line in PowerShell or a command prompt. You can find more information on `http://chocolatey.org` (see Figure 3-1) and you can even create an account and provide your own packages there.




B. Weissman and E. van de Laar, *SQL Server Big Data Clusters*,
https://doi.org/10.1007/978-1-4842-5985-6_3

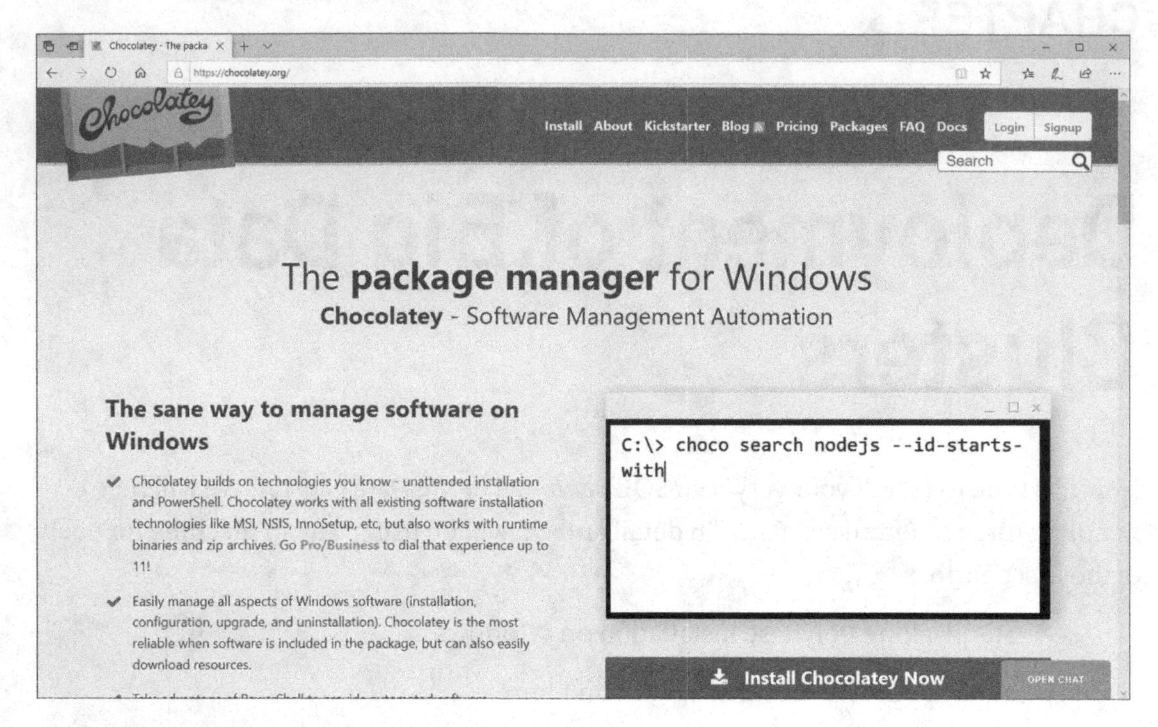


***Figure 3-1.*** *Home page of Chocolatey*

From a simple user perspective though, there is no need to create an account or to download any installer.

To make choco available on your system, open a PowerShell window in Administrative mode and run the script shown in Listing 3-1.

***Listing 3-1.*** Install script for Chocolatey in PowerShell

```
Set-ExecutionPolicy Bypass -Scope Process -Force; [System.Net.ServicePointManager]::SecurityProtocol = [System.Net.ServicePointManager]::SecurityProtocol -bor 3072; iex ((New-Object System.Net.WebClient).DownloadString('https://chocolatey.org/install.ps1'))
```

Once the respective command has completed, choco is installed and ready to be used.

## Installation of an On-Premises PolyBase Instance

In case you're only interested in the data virtualization feature of SQL Server 2019 Big Data Clusters, the installation is actually much easier and lightweight than for a full environment. The PolyBase feature, which enabled the data virtualization feature, can be installed during the regular setup routine of SQL Server 2019 on any platform.

If you want to use Teradata through PolyBase, the C++ Redistributable 2012 is required to actually communicate with our SQL Server instance. Having SQL Server Management Studio (SSMS) may be helpful in either case and is nice to have it installed and ready to replay the examples we are showing throughout this book.

Let's install the packages we mentioned earlier through Chocolatey. Just run the three commands from Listing 3-2 and choco will take care of the rest.

***Listing 3-2.*** Install script for PolyBase prerequisites

```
choco install sql-server-management-studio -y
choco install vcredist2012 -y
```

With our prerequisites installed, we can get to the actual SQL Server installation. Navigate to `www.microsoft.com/en-us/evalcenter/evaluate-sql-server-2019` and follow the instructions to download.

Run the downloaded file, as shown in Figure 3-2.

Select "Download Media" as the installation type.

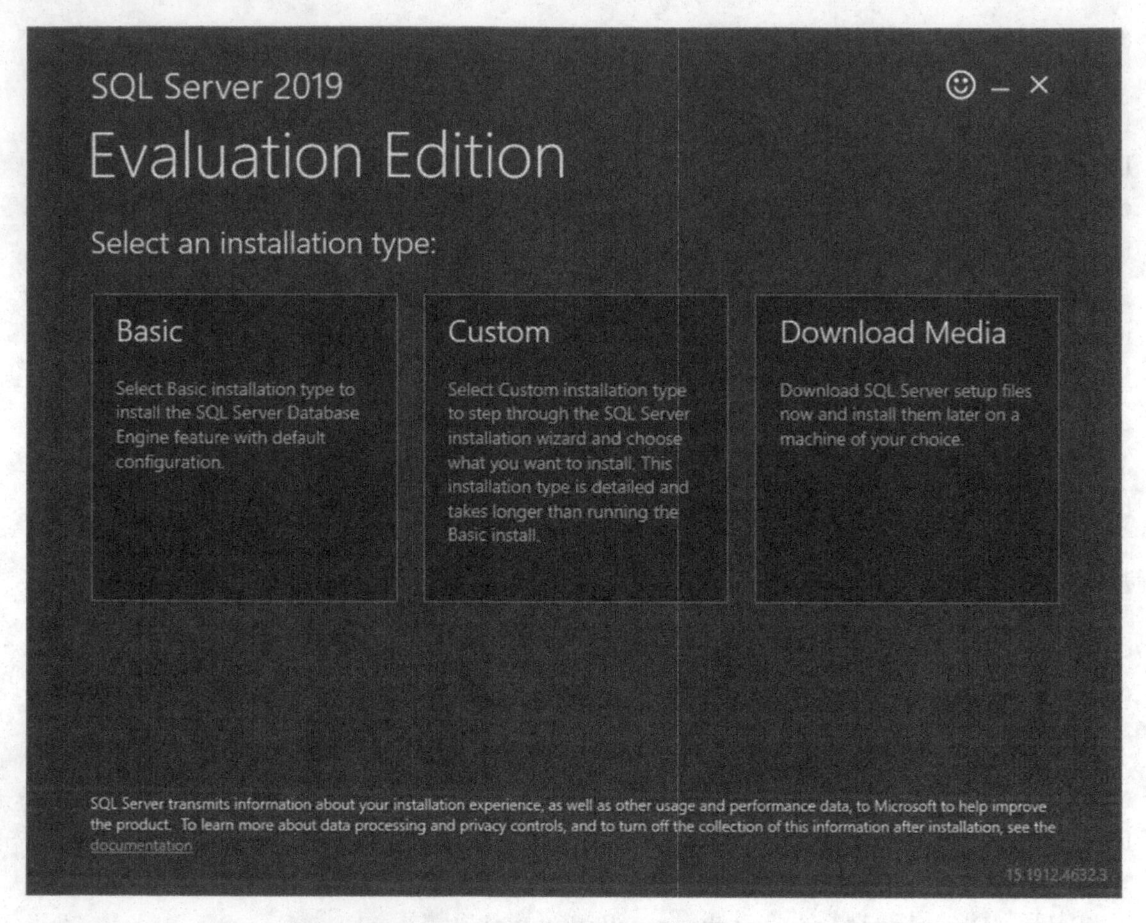


***Figure 3-2.*** *SQL Server 2019 installer – Installation type selection*

Then confirm language and directory as shown in Figure 3-3.

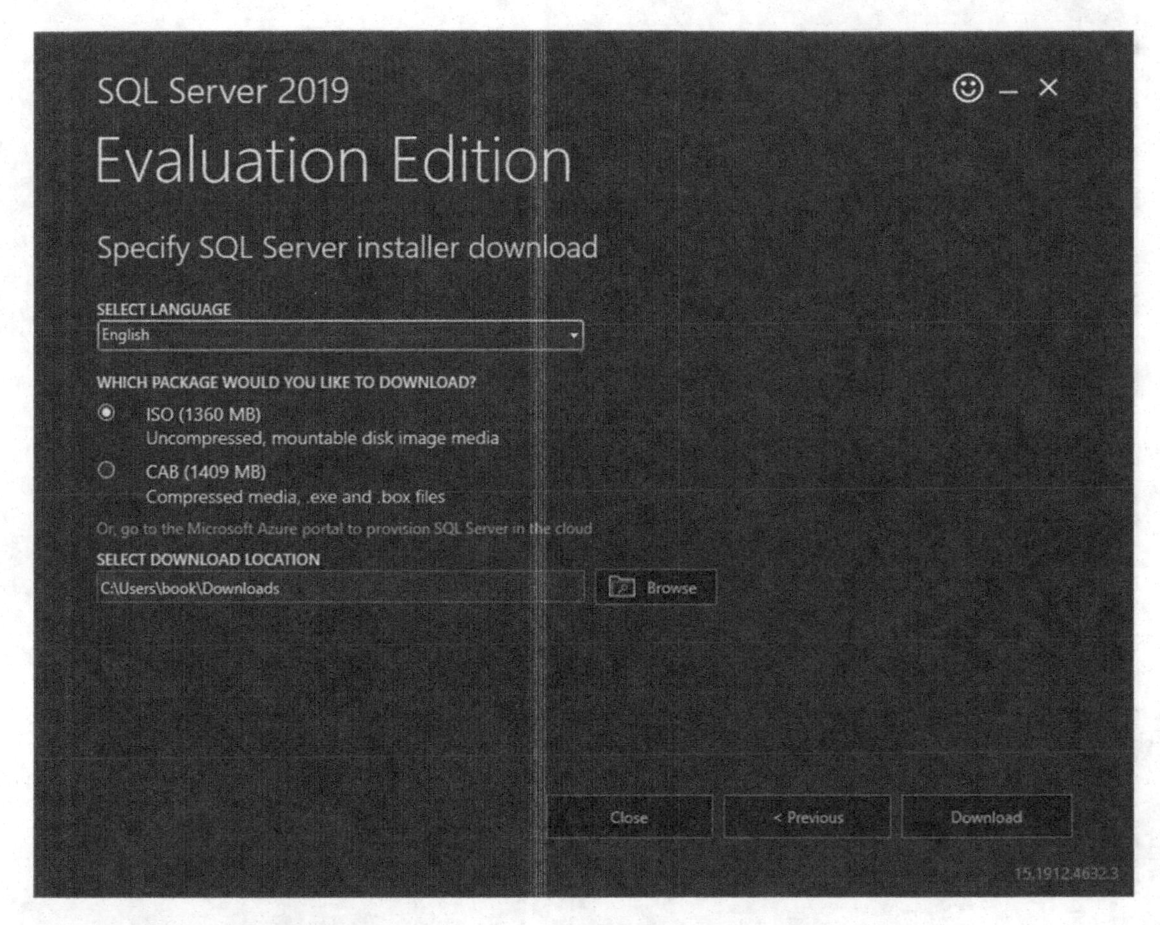


***Figure 3-3.*** *SQL Server 2019 installer – Download Media dialog*

When the download is complete and successful, you will see the message in Figure 3-4.

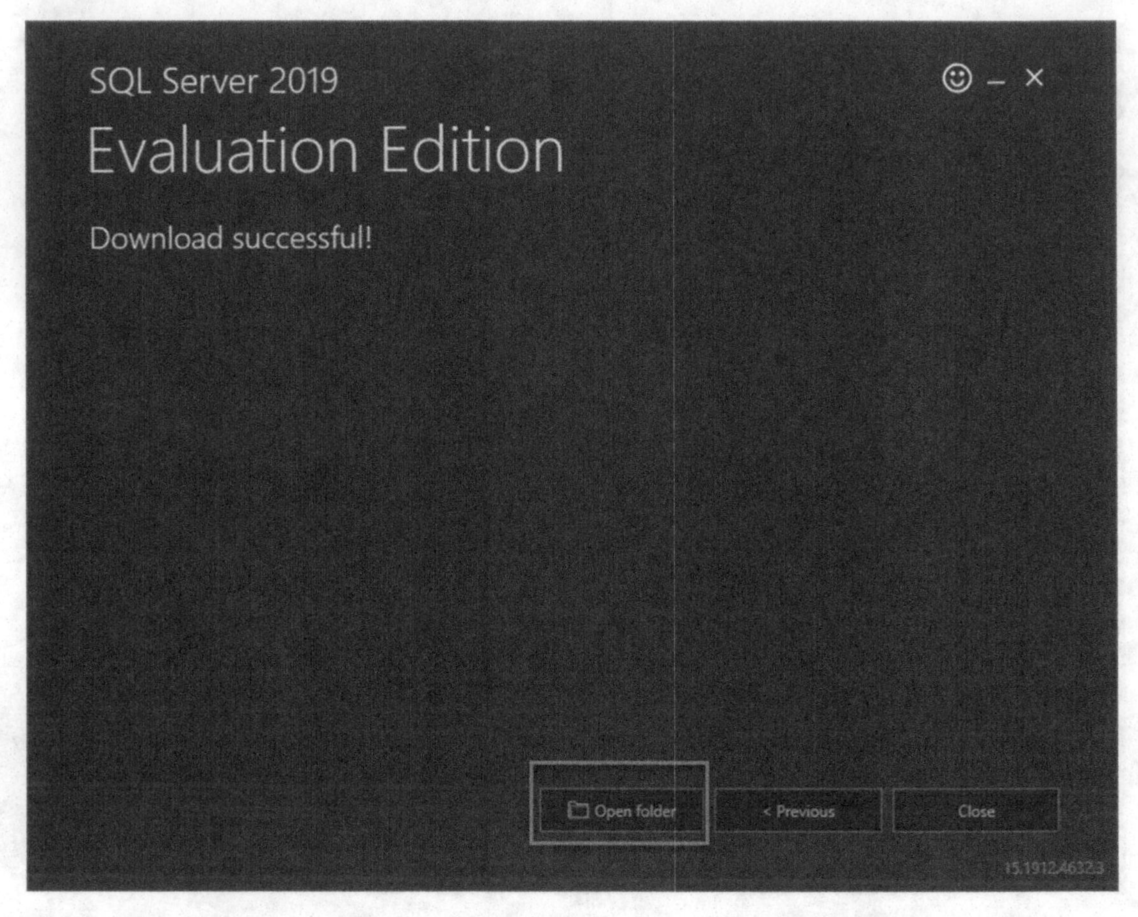


***Figure 3-4.*** *SQL Server 2019 installer – Download Media successful*

Now navigate to the folder in which you have placed the download. Mount the image as shown in Figure 3-5.

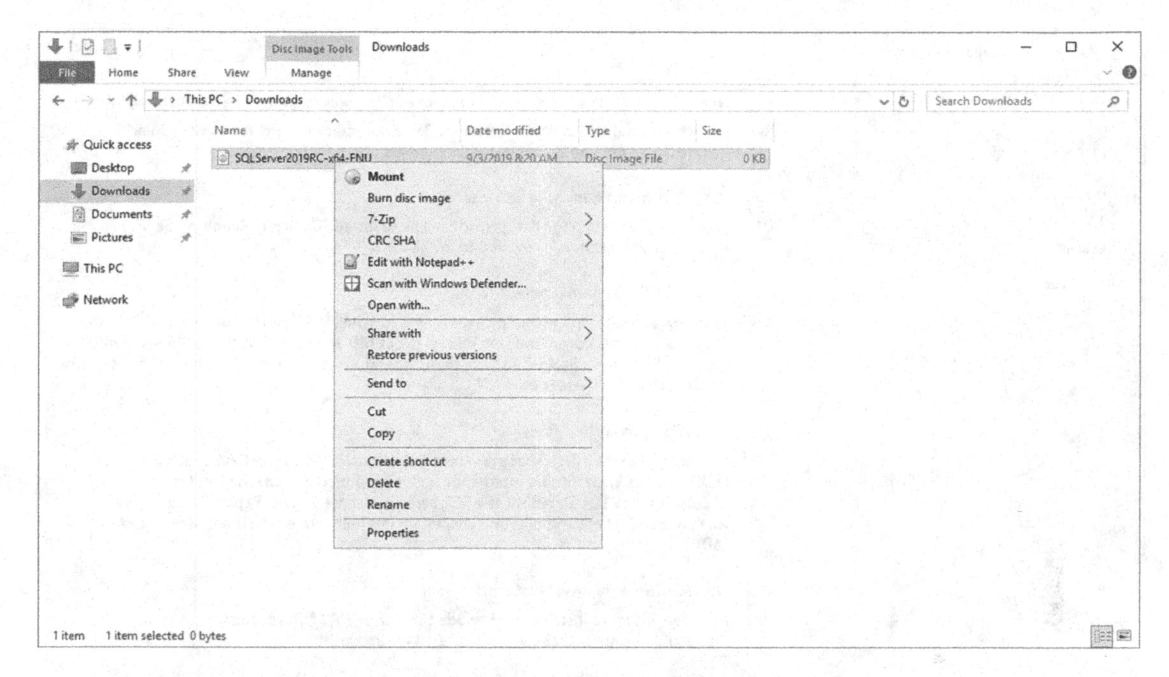


***Figure 3-5.*** *SQL Server 2019 installer – mount ISO*

The installation can be run unattended, but for a first install, it probably makes more sense to explore your options. Run setup.exe, go to the Installation tab, and pick "New SQL Server stand-alone installation," as shown in Figure 3-6.

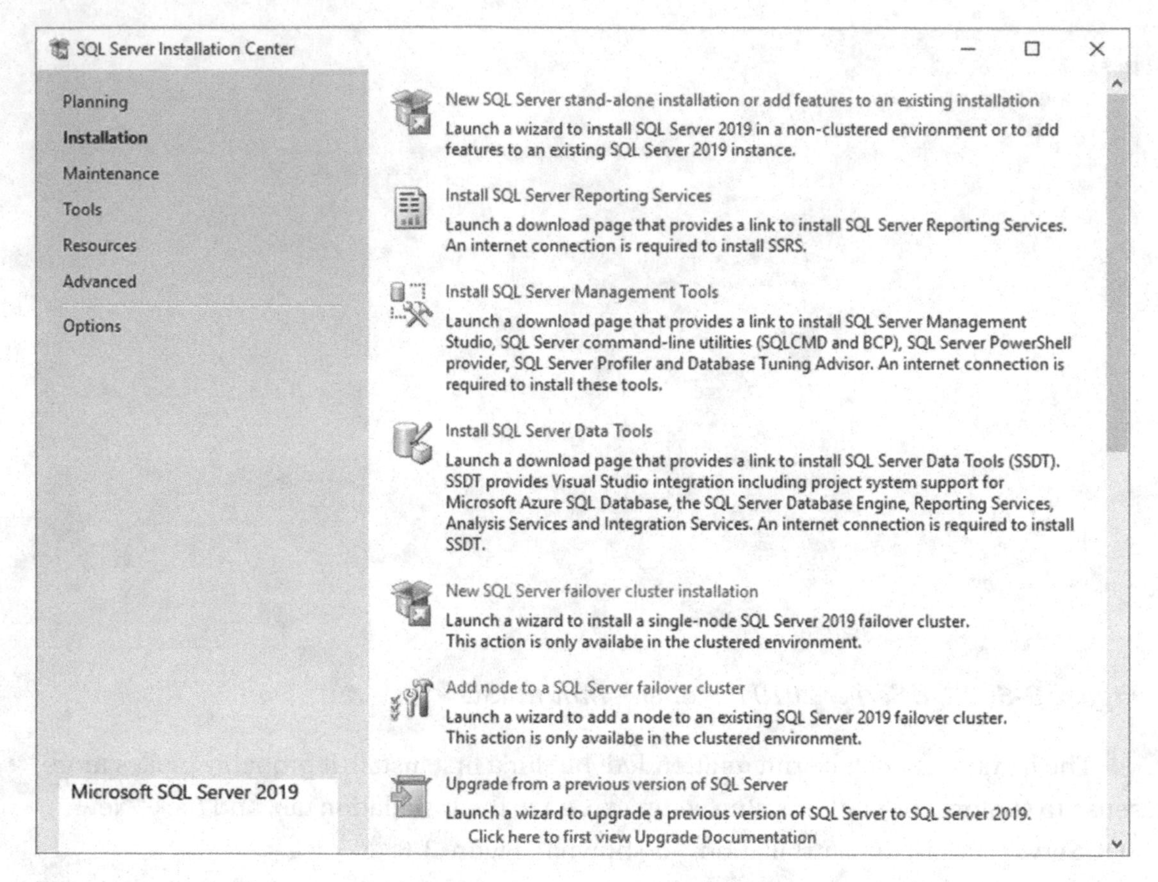


***Figure 3-6.*** *SQL Server 2019 installer – main screen*

Pick the evaluation edition as shown in Figure 3-7, confirm the License Terms, and the check the "Check for updates" check box on the subsequent screens.

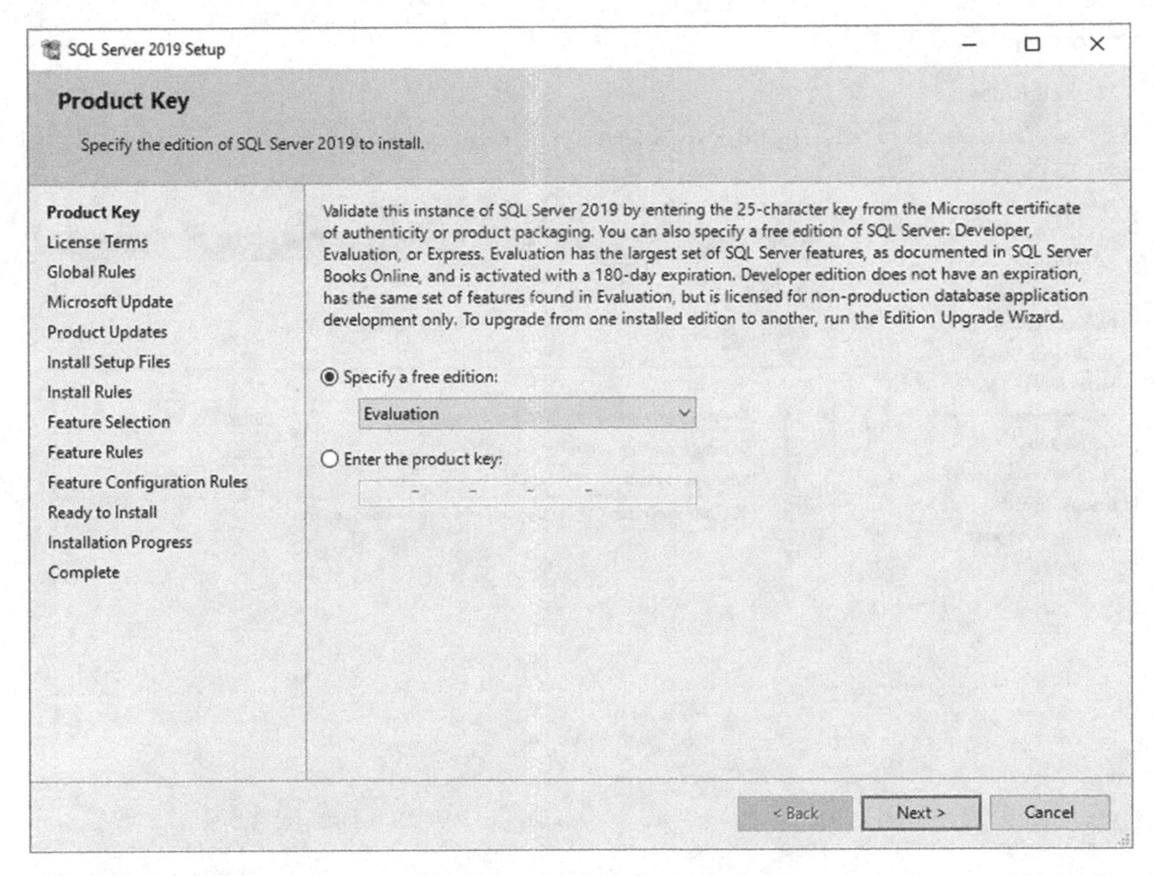


***Figure 3-7.*** *SQL Server 2019 installer – edition selection*

Setup rules identify potential problems that might occur while running the Setup. Failures and warnings as shown in Figure 3-8 must be corrected before the Setup can be completed.

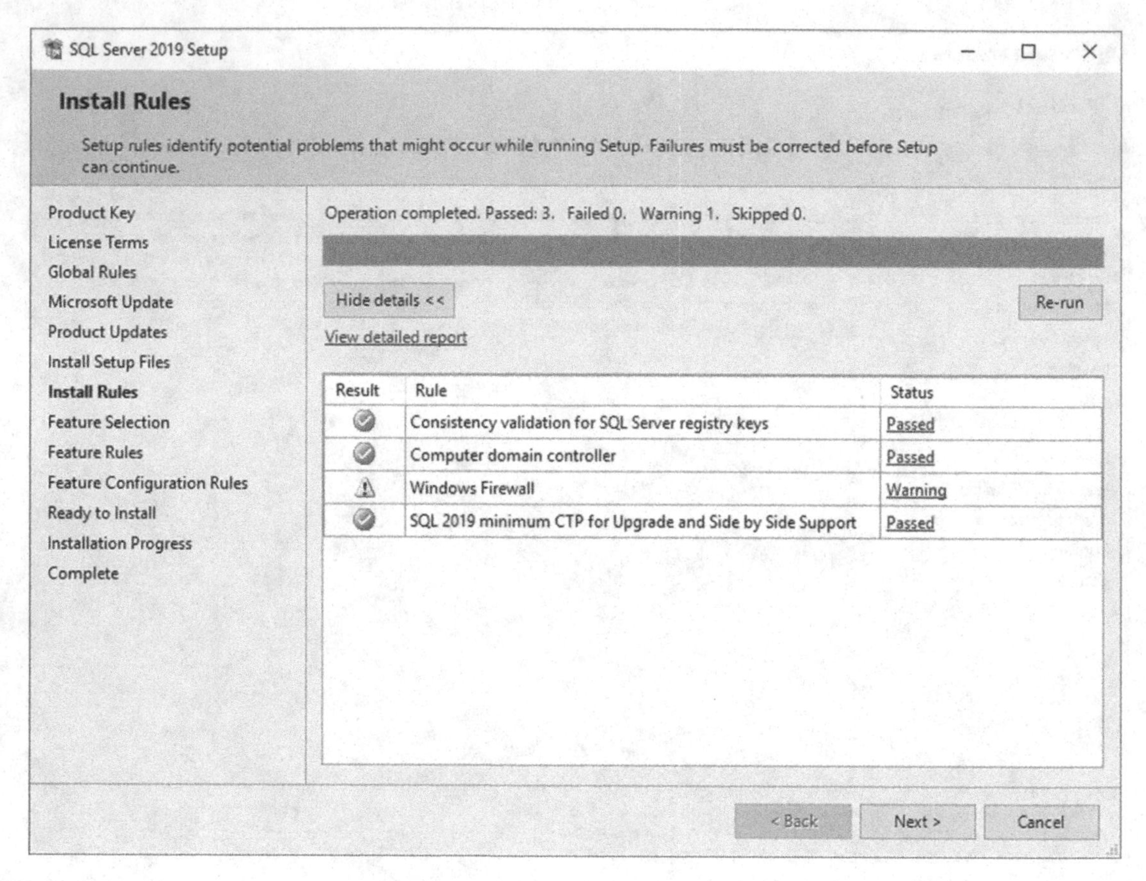


***Figure 3-8.*** *SQL Server 2019 installer – Install Rules*

From the feature selection dialog shown in Figure 3-9, tick the "PolyBase Query Service for External Data." Also tick its child node "Java connector for HDFS data sources."

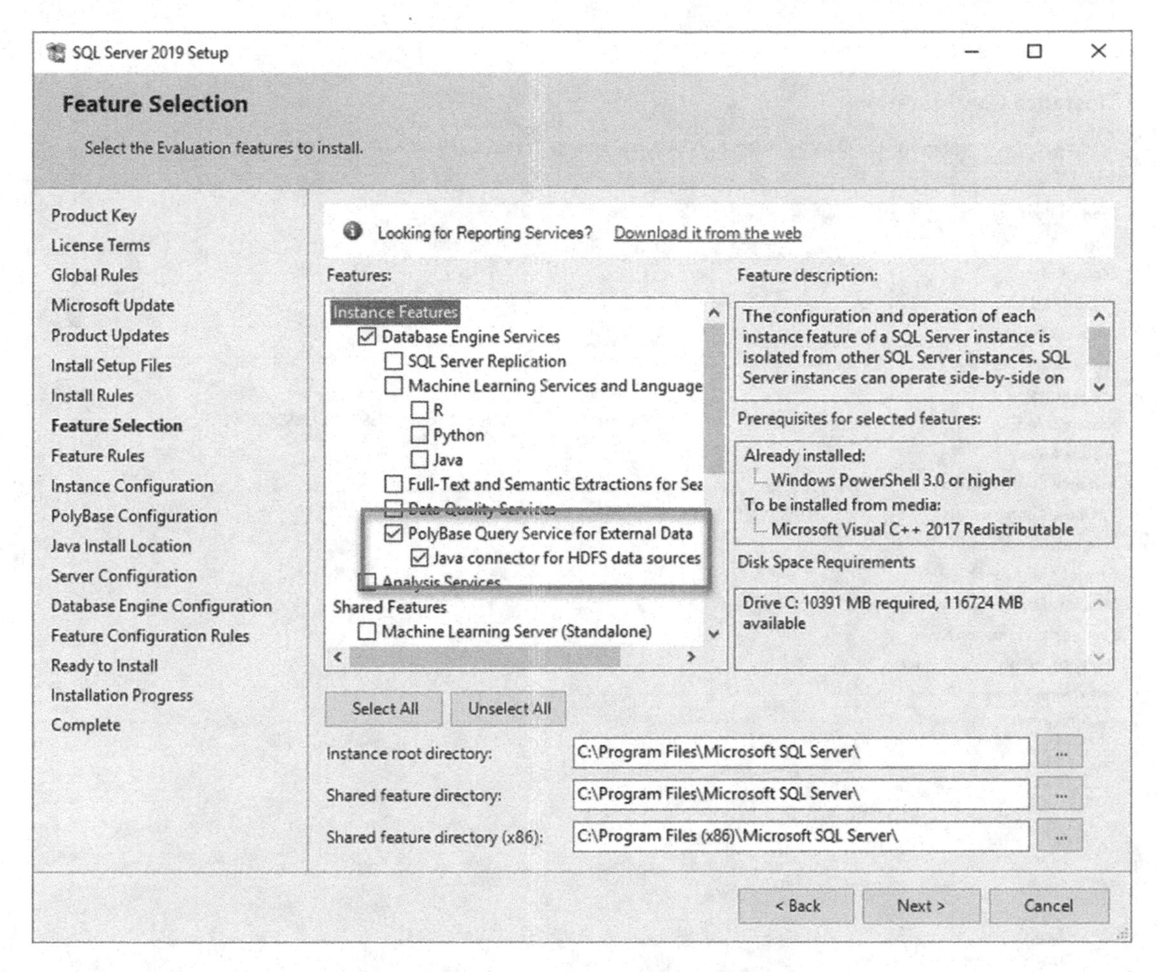


***Figure 3-9.*** *SQL Server 2019 installer – Feature Selection*

Using Instance Configuration specify the name and Instance ID for the Instance SQL Server. The Instance ID as shown in Figure 3-10 becomes part of the installation path.

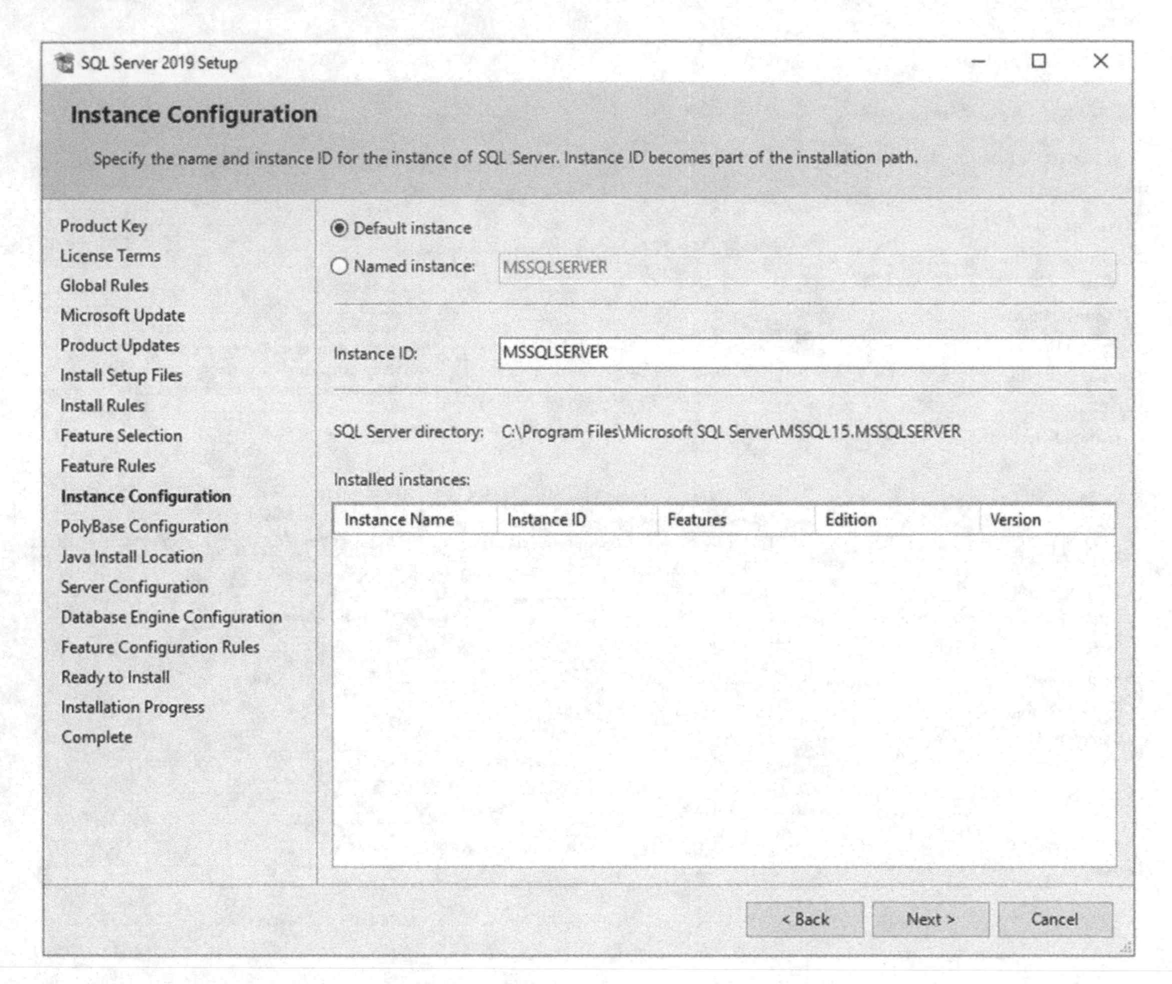


***Figure 3-10.*** *SQL Server 2019 installer – Instance Configuration*

From the dialog in Figure 3-11, choose to configure a stand-alone PolyBase-enabled instance.

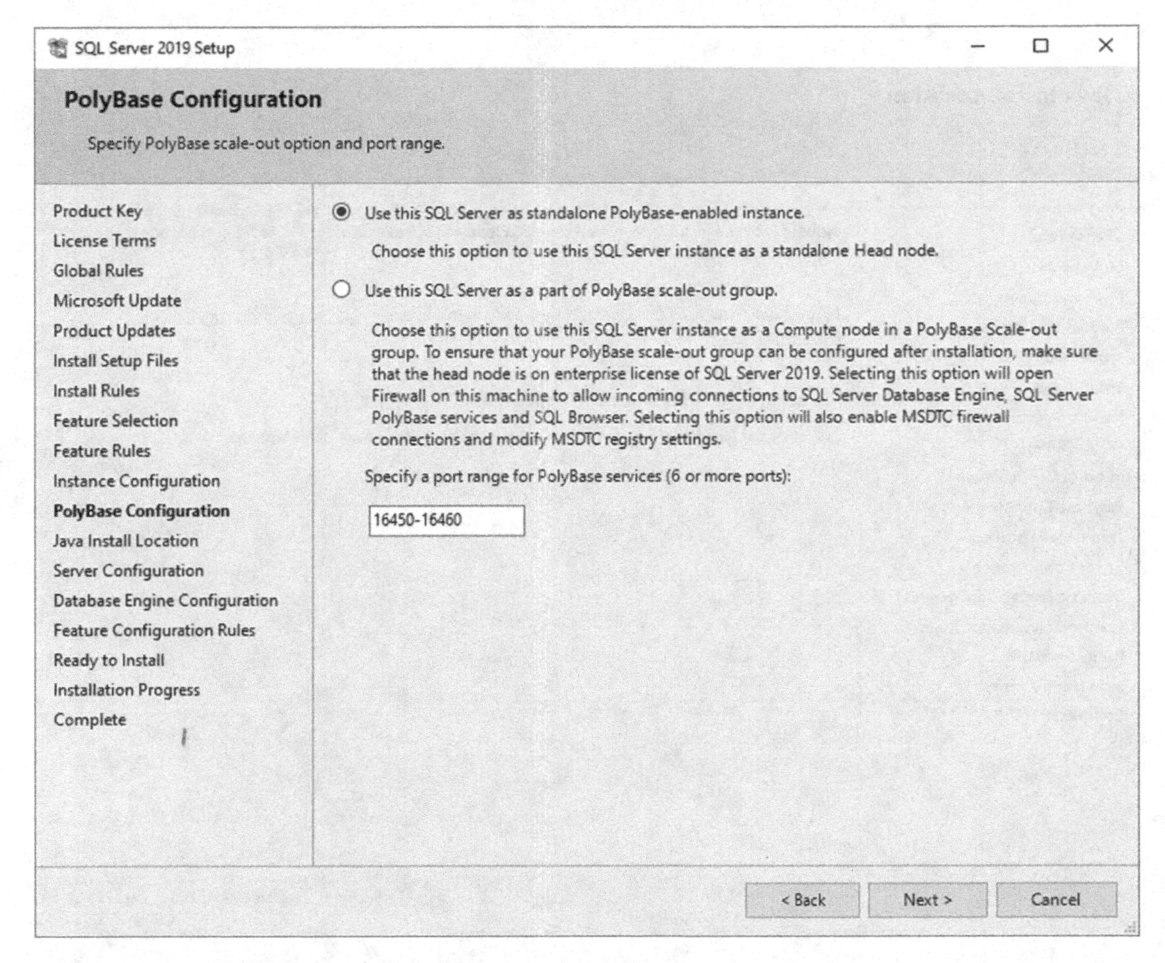


***Figure 3-11.*** *SQL Server 2019 installer – PolyBase Configuration*

As you can see in Figure 3-12, the PolyBase HDFS connector requires Java; you will be prompted to either install Open JRE with SQL Server or provide the location of an existing Java installation on your machine, if there is any.

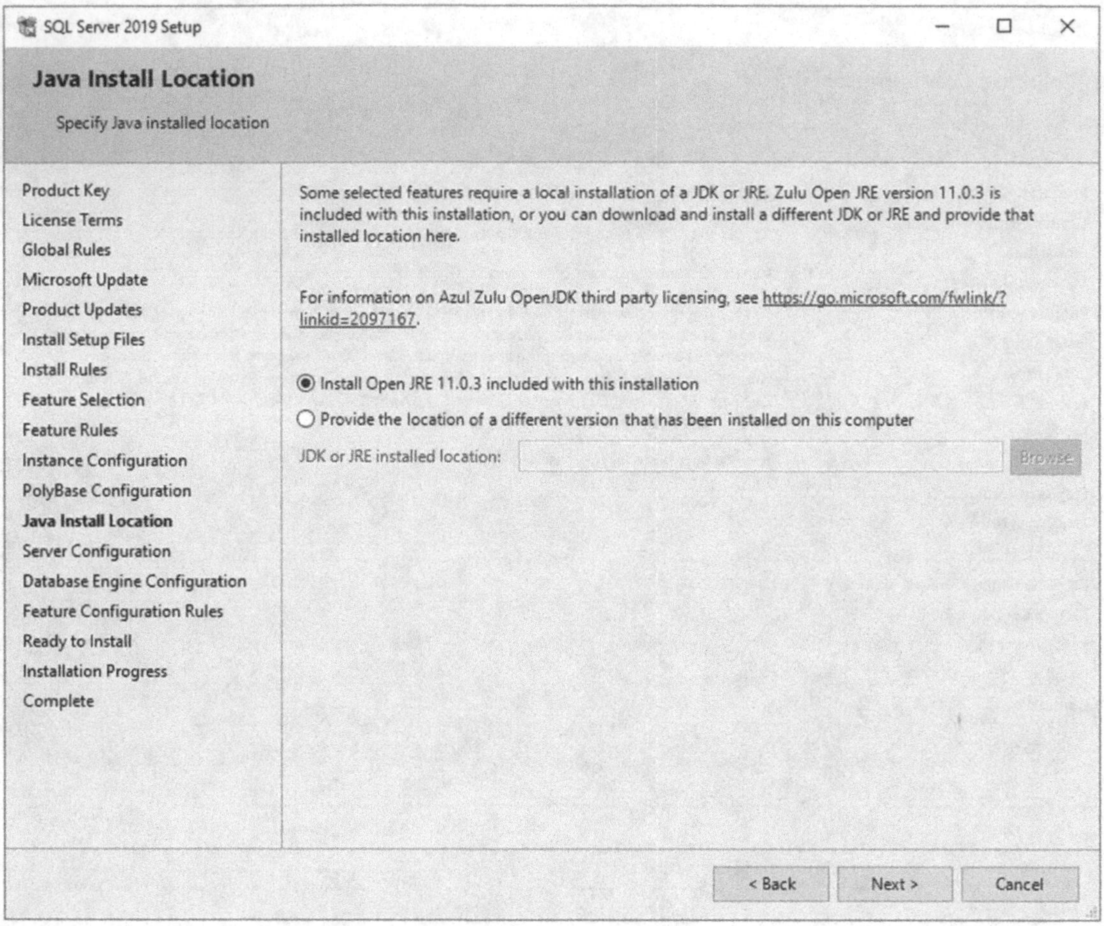

SQL Server 2019 Setup

**Java Install Location**

Specify Java installed location

- Product Key
- License Terms
- Global Rules
- Microsoft Update
- Product Updates
- Install Setup Files
- Install Rules
- Feature Selection
- Feature Rules
- Instance Configuration
- PolyBase Configuration
- **Java Install Location**
- Server Configuration
- Database Engine Configuration
- Feature Configuration Rules
- Ready to Install
- Installation Progress
- Complete

Some selected features require a local installation of a JDK or JRE. Zulu Open JRE version 11.0.3 is included with this installation, or you can download and install a different JDK or JRE and provide that installed location here.

For information on Azul Zulu OpenJDK third party licensing, see https://go.microsoft.com/fwlink/?linkid=2097167.

(•) Install Open JRE 11.0.3 included with this installation

( ) Provide the location of a different version that has been installed on this computer

JDK or JRE installed location: Browse

< Back | Next > | Cancel

***Figure 3-12.*** *SQL Server 2019 installer – Java Install Location*

Then confirm the default accounts as shown in Figure 3-13.

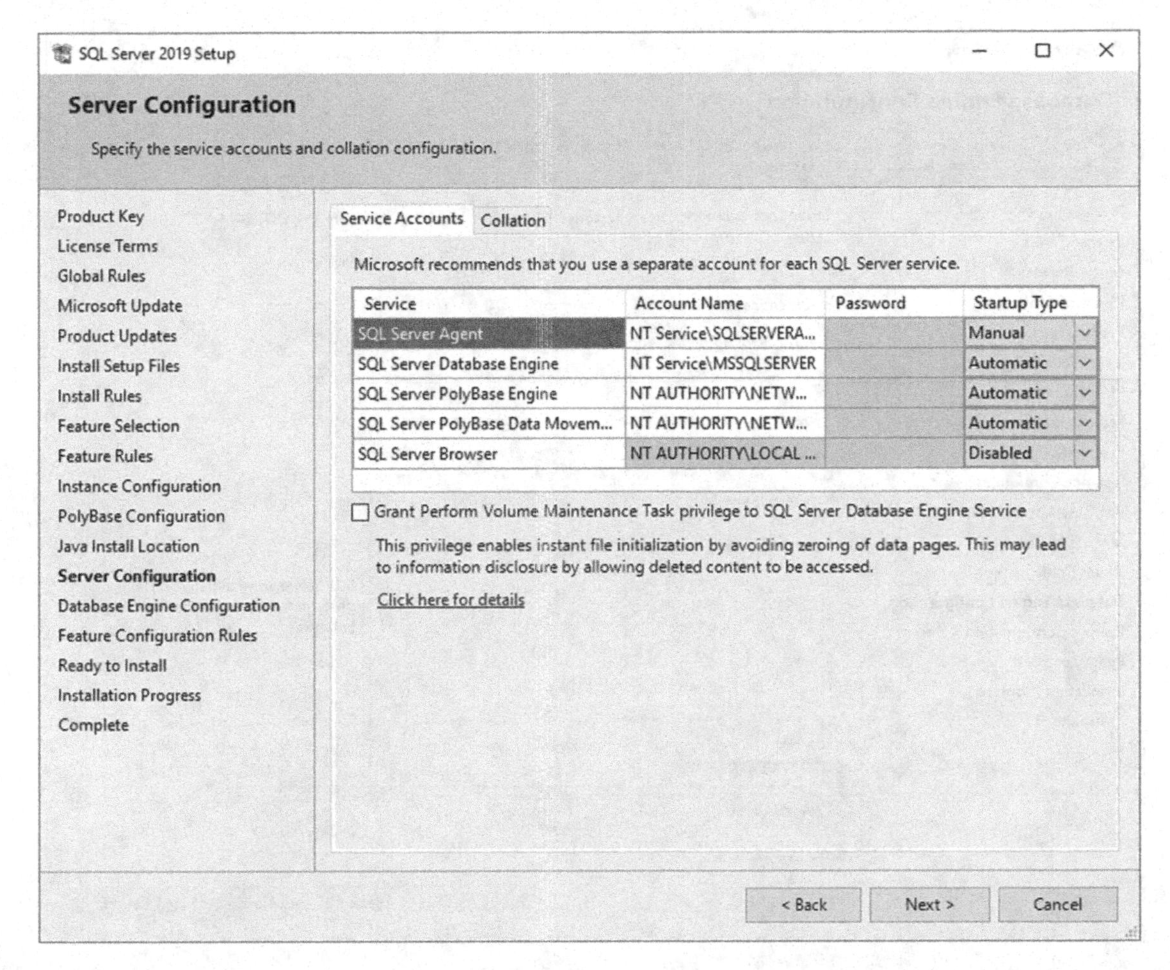


***Figure 3-13.*** *SQL Server 2019 installer – Server Configuration*

Stick with Windows authentication and add your current user as shown in Figure 3-14.

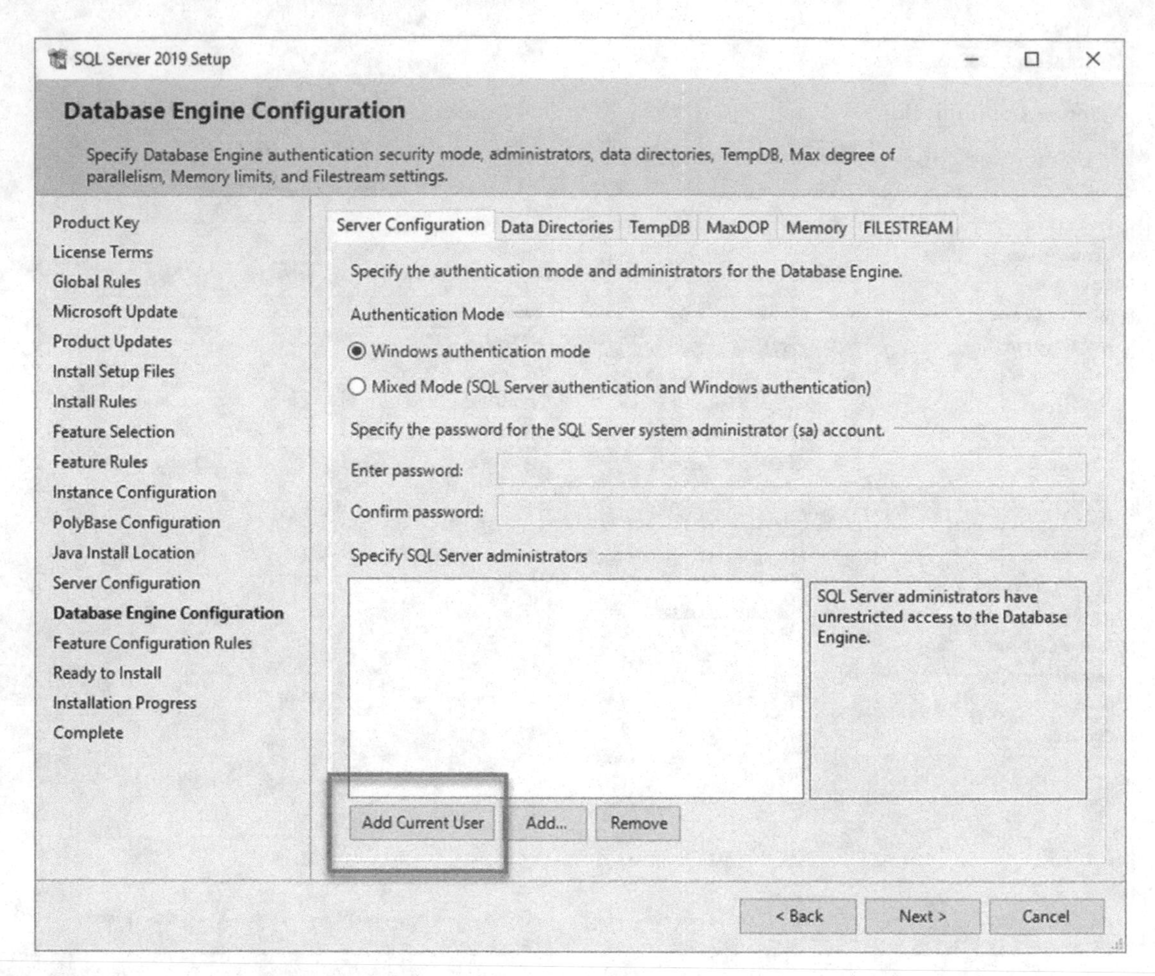


***Figure 3-14.*** *SQL Server 2019 installer – Database Engine Configuration*

Click Install on the summary page shown in Figure 3-15 and wait for the installer to finish.

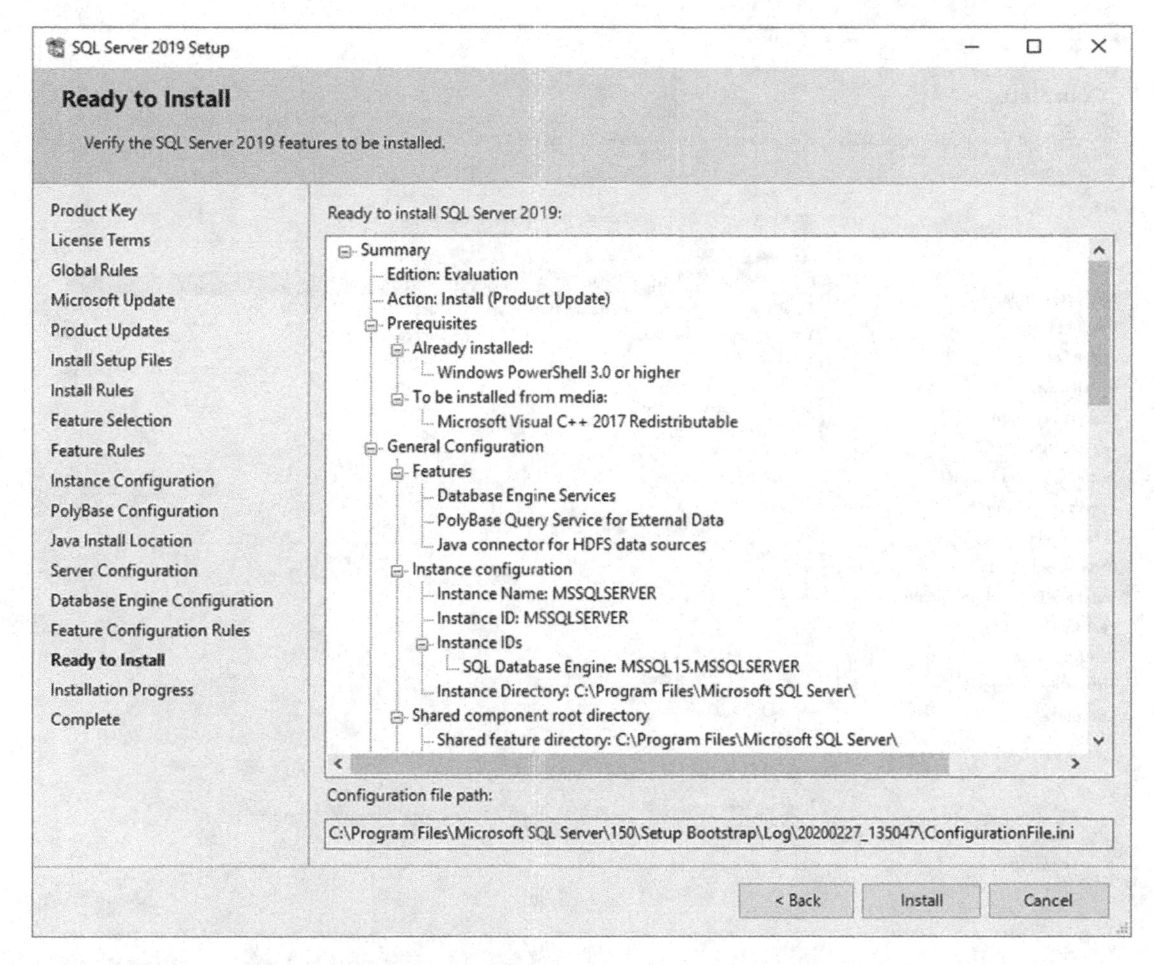


***Figure 3-15.*** *SQL Server 2019 installer – overview*

Once the setup is successfully completed, a status summary as shown in Figure 3-16 is displayed and you can close the installer.

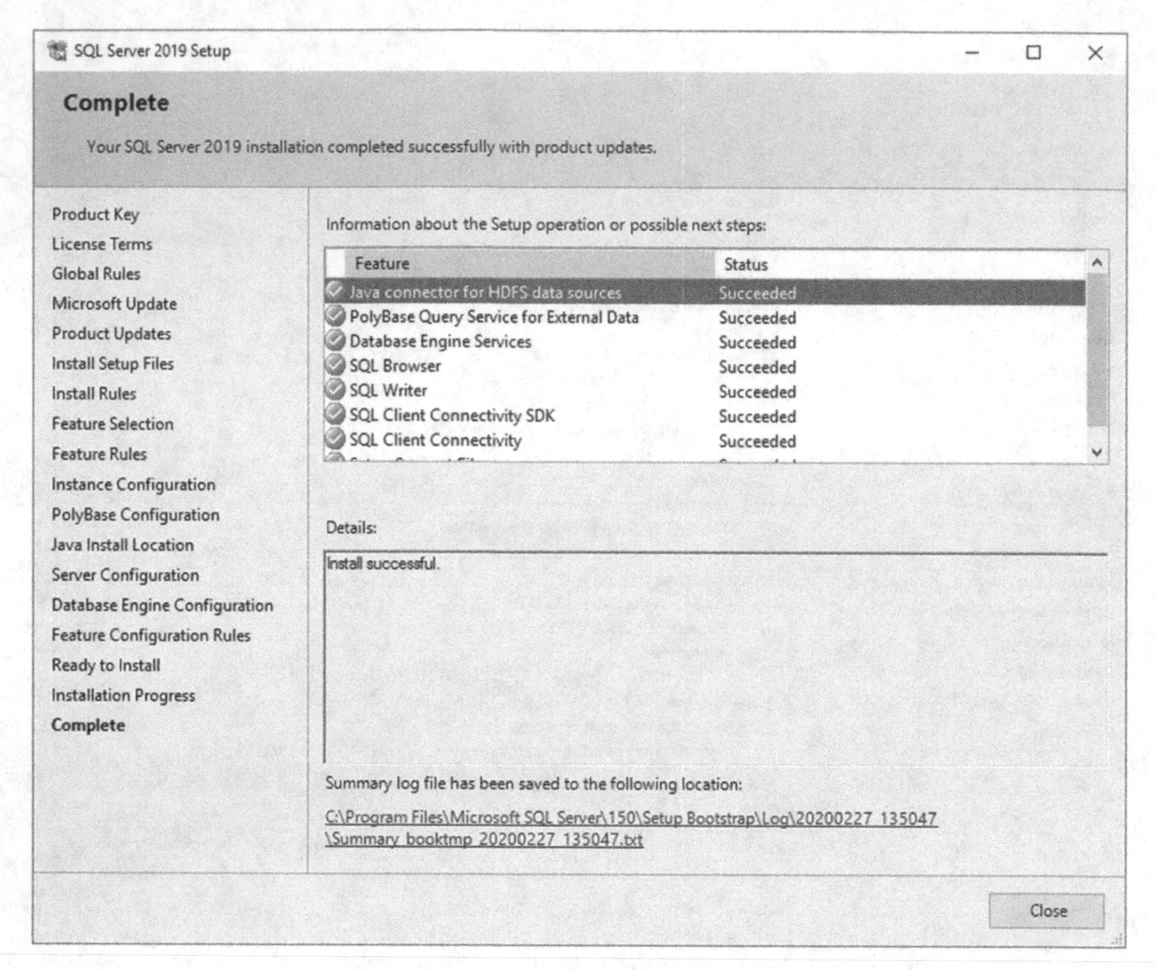


***Figure 3-16.*** *SQL Server 2019 installer – Complete*

Connect to the instance using SQL Server Management Studio (SSMS) or any other SQL Server client tool like Azure Data Studio, open a new query, and run the script shown in Listing 3-3.

***Listing 3-3.*** Enable PolyBase through T-SQL

```
exec sp_configure @configname = 'polybase enabled', @configvalue = 1;
RECONFIGURE
```

The output should be

```
Configuration option 'polybase enabled' changed from 0 to 1. Run the
RECONFIGURE statement to install.
```

Click "Restart" in the Object Explorer menu as shown in Figure 3-17 to restart the SQL Server Instance.

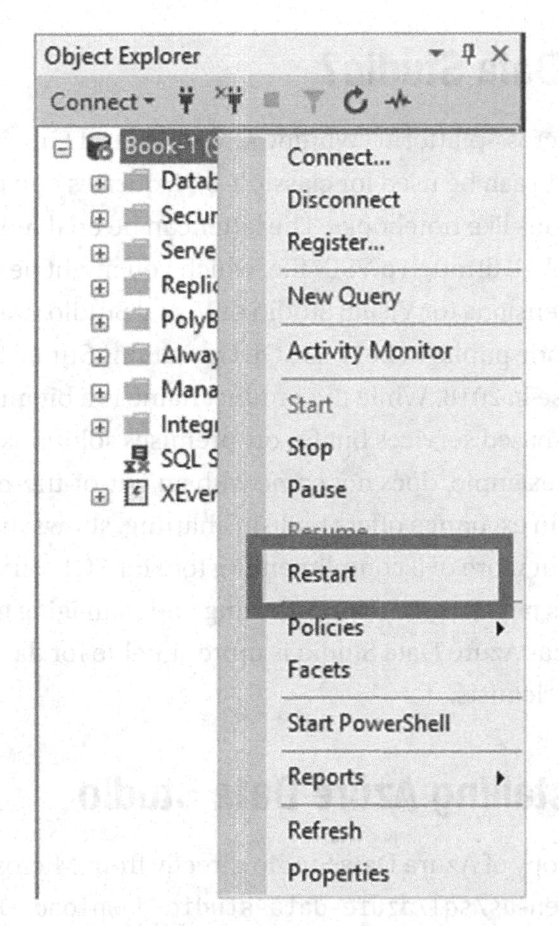


***Figure 3-17.*** *Restart SQL Server Instance*

You're done – you do now have access to a PolyBase-enabled SQL Server 2019 installation.

# Using Azure Data Studio to Work with Big Data Clusters

As part of Microsoft's SQL Client tool strategy, it may not surprise you that most of the tasks necessary to work with a Big Data Cluster are achieved through Azure Data Studio (ADS) rather than SQL Server Management Studio or other tools. For those of you who are not familiar with this tool, we're going to start with a little introduction including how to get your hands on this tool.

## What Is Azure Data Studio?

Azure Data Studio is a cross-platform (Windows, MacOS, and Linux), extendable, and customizable tool which can be used for classic T-SQL queries and commands, as well as multiple new functions like notebooks. The latter can be enabled through extensions which are usually installed through a VSIX file, which you might be familiar with from working with other extensions for Visual Studio or Visual Studio Code.[1]

It was originally made public in 2017 as SQL Operations Studio but was rebranded before its official release in 2018. While the product name is a bit misleading, it is not only for cloud (Azure)-based services but for on-premises solutions and needs as well.

The fact that it, for example, does not come with an out-of-the-box interface for SQL Server Agent, but in exchange offers built-in charting, shows that it is not so much a replacement but more of a complimenting tool for SQL Server Management Studio (SSMS). SSMS is targeting an administrating and managing group (database administrators), whereas Azure Data Studio is more suitable for data professionals of all kinds, including data scientists.

## Getting and Installing Azure Data Studio

You can get your free copy of Azure Data Studio directly from Microsoft at `https://docs.microsoft.com/en-us/sql/azure-data-studio/download`. Download the installer of your choice for your platform and run it.

Alternatively, simply run this Chocolatey command (Listing 3-4) which will install the latest version for you.

***Listing 3-4.*** Install ADS via choco

```
choco install azure-data-studio -y
```

[1] https://docs.microsoft.com/en-us/visualstudio/extensibility/shipping-visual-studio-extensions?view=vs-2019

# Installation of a "Real" Big Data Cluster

If you want to make use of the full Big Data Cluster feature set, you will need a full installation including all the different roles and pools.

## kubeadm on Linux

A very easy way to deploy a Big Data Cluster is using kubeadm on a vanilla (freshly installed) Ubuntu 16.04 or 18.04 virtual or physical machine.

Microsoft provides a script for you that does all the work, so besides the Linux installation itself, there is not much to do for you, which makes this probably the easiest way to get your first Big Data Cluster up and running.

First, make sure your Linux machine is up to date by running the commands in Listing 3-5.

***Listing 3-5.*** Patch Ubuntu

```
sudo apt update&&apt upgrade -y
sudo systemctl reboot
```

Then, download the script, make it executable, and run it with root permissions as shown in Listing 3-6.

***Listing 3-6.*** Download and execute the deployment script

```
curl --output setup-bdc.sh https://raw.githubusercontent.com/microsoft/
sql-server-samples/master/samples/features/sql-big-data-cluster/deployment/
kubeadm/ubuntu-single-node-vm/setup-bdc.sh
chmod +x setup-bdc.sh
sudo ./setup-bdc.sh
```

As you can see in Figure 3-18, the script will ask you for a password and automatically start preparational steps afterward.

```
bigdata@bdc-ubuntu:~$ sudo ./setup-bdc.sh
Create Password for Big Data Cluster:
Confirm your Password:

##############################################################################
Starting installing packages...
Hit:1 http://azure.archive.ubuntu.com/ubuntu bionic InRelease
Get:2 http://azure.archive.ubuntu.com/ubuntu bionic-updates InRelease [88.7 kB]
Get:3 http://azure.archive.ubuntu.com/ubuntu bionic-backports InRelease [74.6 kB]
Get:4 http://security.ubuntu.com/ubuntu bionic-security InRelease [88.7 kB]
Get:5 http://azure.archive.ubuntu.com/ubuntu bionic-updates/main amd64 Packages [722 kB]
Get:6 http://azure.archive.ubuntu.com/ubuntu bionic-updates/main Translation-en [262 kB]
Get:7 http://azure.archive.ubuntu.com/ubuntu bionic-updates/restricted amd64 Packages [13.1 kB]
Get:8 http://azure.archive.ubuntu.com/ubuntu bionic-updates/restricted Translation-en [4448 B]
Get:9 http://azure.archive.ubuntu.com/ubuntu bionic-updates/universe amd64 Packages [1003 kB]
Get:10 http://azure.archive.ubuntu.com/ubuntu bionic-updates/universe Translation-en [308 kB]
Get:11 http://azure.archive.ubuntu.com/ubuntu bionic-updates/multiverse amd64 Packages [7308 B]
Get:12 http://azure.archive.ubuntu.com/ubuntu bionic-updates/multiverse Translation-en [3836 B]
Get:13 http://azure.archive.ubuntu.com/ubuntu bionic-backports/universe amd64 Packages [4000 B]
Get:14 http://azure.archive.ubuntu.com/ubuntu bionic-backports/universe Translation-en [1856 B]
Get:15 http://security.ubuntu.com/ubuntu bionic-security/main amd64 Packages [489 kB]
Get:16 http://security.ubuntu.com/ubuntu bionic-security/main Translation-en [166 kB]
Get:17 http://security.ubuntu.com/ubuntu bionic-security/restricted amd64 Packages [4976 B]
Get:18 http://security.ubuntu.com/ubuntu bionic-security/restricted Translation-en [2476 B]
Get:19 http://security.ubuntu.com/ubuntu bionic-security/universe amd64 Packages [600 kB]
Get:20 http://security.ubuntu.com/ubuntu bionic-security/universe Translation-en [200 kB]
```

***Figure 3-18.*** *Deployment on Linux with kubeadm*

After pre-fetching the images, provisioning Kubernetes, and all other required steps, the deployment of the Big Data Cluster is started as shown in Figure 3-19.

```
################################################################################
Starting to deploy azdata cluster...
The privacy statement can be viewed at:
https://go.microsoft.com/fwlink/?LinkId=853010

The license terms for azdata can be viewed at:
https://aka.ms/azdata-eula

kubeadm-custom/bdc.json created
kubeadm-custom/control.json created
The privacy statement can be viewed at:
https://go.microsoft.com/fwlink/?LinkId=853010

The license terms for SQL Server Big Data Cluster can be viewed at:
https://go.microsoft.com/fwlink/?LinkId=2002534


Cluster deployment documentation can be viewed at:
https://aka.ms/bdc-deploy


NOTE: Cluster creation can take a significant amount of time depending on
configuration, network speed, and the number of nodes in the cluster.

Starting cluster deployment.
Waiting for cluster controller to start.
Cluster controller endpoint is available at 10.0.1.4:30080.
```

***Figure 3-19.*** *Deployment on Linux with kubeadm*

Once the whole script completes, you are done! As demonstrated in Figure 3-20, the script will also provide all the endpoints that were created during the deployment.

```
Cluster control plane is ready.
Data pool is ready.
Compute pool is ready.
Storage pool is ready.
Master pool is ready.
Cluster deployed successfully.
Azdata cluster created.
Context "kubernetes-admin@kubernetes" modified.
Logged in successfully to `https://10.0.1.4:30080`
Description                                            Endpoint                                                 Name               Protocol
-----------------------------------------------------  -------------------------------------------------------  -----------------  ----------
Gateway to access HDFS files, Spark                    https://10.0.1.4:30443                                   gateway            https
Spark Jobs Management and Monitoring Dashboard         https://10.0.1.4:30443/gateway/default/sparkhistory      spark-history      https
Spark Diagnostics and Monitoring Dashboard             https://10.0.1.4:30443/gateway/default/yarn              yarn-ui            https
Application Proxy                                      https://10.0.1.4:30778                                   app-proxy          https
Management Proxy                                       https://10.0.1.4:30777                                   mgmtproxy          https
Log Search Dashboard                                   https://10.0.1.4:30777/kibana                            logsui             https
Metrics Dashboard                                      https://10.0.1.4:30777/grafana                           metricsui          https
Cluster Management Service                             https://10.0.1.4:30080                                   controller         https
SQL Server Master Instance Front-End                   10.0.1.4,31433                                           sql-server-master  tds
HDFS File System Proxy                                 https://10.0.1.4:30443/gateway/default/webhdfs/v1        webhdfs            https
Proxy for running Spark statements, jobs, applications https://10.0.1.4:30443/gateway/default/livy/v1           livy               https
```

***Figure 3-20.*** *Successful deployment on Linux with kubeadm*

Your cluster is now fully deployed and ready!

## Azure Kubernetes Service (AKS)

Another straightforward way to deploy your cluster is to use Azure Kubernetes Service, in which the Kubernetes cluster is set up and provided in the Microsoft Azure cloud. While the deployment is started and controlled through any machine (either your local PC or a VM), the cluster itself will run in Azure, so this means that deployment requires an Azure account and will result in cost on your Azure subscription.

You can deploy either through a wizard in Azure Data Studio or manually through a tool called azdata (which was also called by the script deploying your previous cluster on Linux). Both methods have some prerequisites that need to be installed first. A full installation actually requires several tools and helpers:

- Python
- Curl and the SQL Server command-line utilities

  So we can communicate with the cluster and upload data to it.
- The Kubernetes CLI
- azdata

  This will create, maintain, and delete a big data cluster.
- Notepad++ and 7Zip

  These are not actual requirements, but if you want to debug your installation, you will get a tar.gz file with potentially huge text files. Windows does not handle these out of the box.

The script in Listing 3-7 will install those to your local machine (or whichever machine you are running the script on), as this is where the deployment is controlled and triggered from. We will be installing those prerequisites through Chocolatey.

***Listing 3-7.*** Install script for Big Data Cluster prerequisites

```
choco install python3 -y
choco install sqlserver-cmdlineutils -y
$env:Path = [System.Environment]::GetEnvironmentVariable("Path","Machine")
+ ";" + [System.Environment]::GetEnvironmentVariable("Path","User")
python -m pip install --upgrade pip
python -m pip install requests
```

```
python -m pip install requests --upgrade
choco install curl -y
choco install kubernetes-cli -y
choco install notepadplusplus -y
choco install 7zip -y
choco install visualcpp-build-tools -y
pip3 install kubernetes
pip3 install -r https://aka.ms/azdata
```

While the respective vendors obviously supply visual/manual installation routines for most of these tools, the scripted approach just makes the whole experience a lot easier.

In addition, as we want to deploy to Azure using a script, we need the azure-cli package shown in Listing 3-8 to be able to connect to our Azure subscription.

***Listing 3-8.*** Install azure-cli

```
choco install azure-cli -y
```

While you technically could prepare everything (we need a resource group, the Kubernetes cluster, etc.) in the Azure Portal or through manual PowerShell scripts, there is a much easier way: get the Python script from `https://github.com/Microsoft/sql-server-samples/tree/master/samples/features/sql-big-data-cluster/deployment/aks`, which will automatically take care of the whole process and setup for you.

Download the script to your desktop or another suitable location and open a command prompt. Navigate to the folder where you've saved the script. You can also download using a command prompt as shown in Listing 3-9.

***Listing 3-9.*** Download deployment script

```
curl --output deploy-sql-big-data-aks.py https://raw.githubusercontent.com/
microsoft/sql-server-samples/master/samples/features/sql-big-data-cluster/
deployment/aks/deploy-sql-big-data-aks.py
```

Of course, you can also modify and review the script as per your needs, for example, to make some parameters like the VM size static rather than a variable or to change the defaults for some of the values.

First, we need to log on to Azure which will be done with the command shown in Listing 3-10.

***Listing 3-10.*** Trigger login to azure from command prompt

```
az login
```

A website will open; log on using your Azure credentials as shown in Figure 3-21.

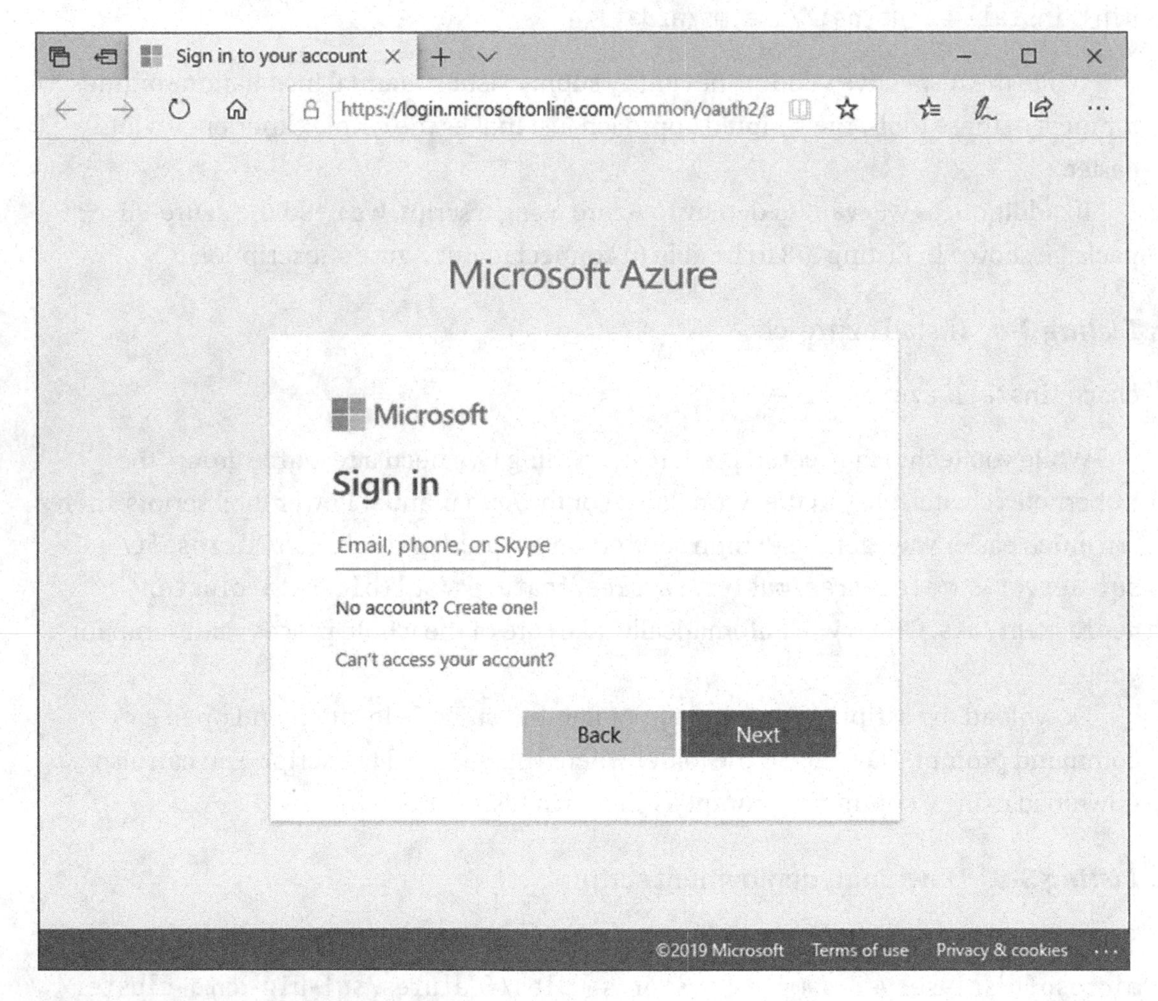


***Figure 3-21.*** *Azure logon screen*

The website will confirm that you are logged on; you can close the browser as shown in Figure 3-22.

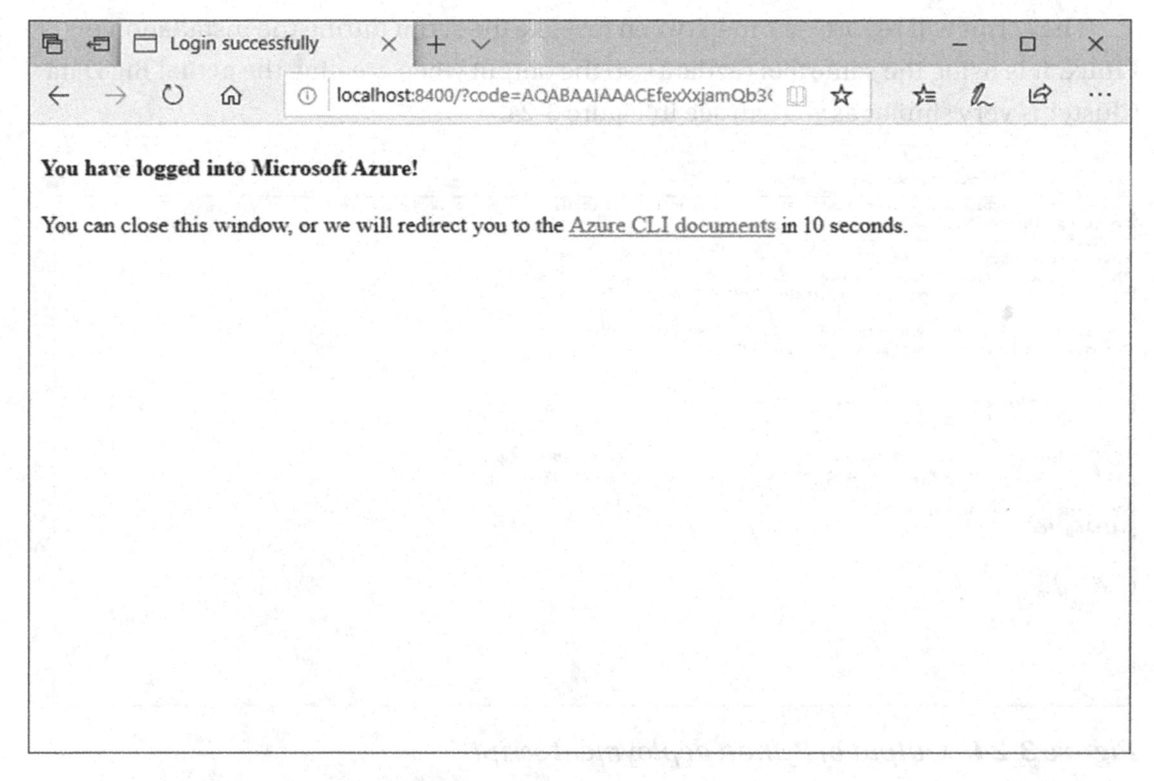


*Figure 3-22. Azure logon confirmation*

Your command prompt, as shown in Figure 3-23, shows all subscriptions linked to the credentials you just used. Copy the ID of the subscription you want to use and execute the Python script, which will ask for everything ranging from subscription ID to the number of nodes inside the Kubernetes cluster.

```
Select Administrator: C:\Windows\System32\cmd.exe - .\deploy-sql-big-data-aks.py

C:\Users\bdcbook\Desktop>.\deploy-sql-big-data-aks.py
Provide your Azure subscription ID:<myID>
Provide Azure resource group name to be created:<myRG>
Provide Azure region - Press ENTER for using `westus`:
Provide VM size for the AKS cluster - Press ENTER for using  `Standard_L8s`:
Provide number of worker nodes for AKS cluster - Press ENTER for using  `1`:
Provide name of AKS cluster and SQL big data cluster - Press ENTER for using  `sqlbigdata`:
Provide username to be used for Controller and SQL Server master accounts - Press ENTER for using  `admin`:
Provide password to be used for Controller user, Knox user (root) and SQL Server Master accounts - Press ENTER for using
 `MySQLBigData2019`_
```

*Figure 3-23. Input of parameters in Python deployment script*

The script now runs through all the required steps. Again, this can take from a couple of minutes to hours, depending on the size of VM, number of nodes, and so on.

The script will report back in between just like the script during the installation on Linux. It is using the same tool (azdata), so the output when creating the actual Big Data Cluster is very similar as you can see in Figure 3-24.

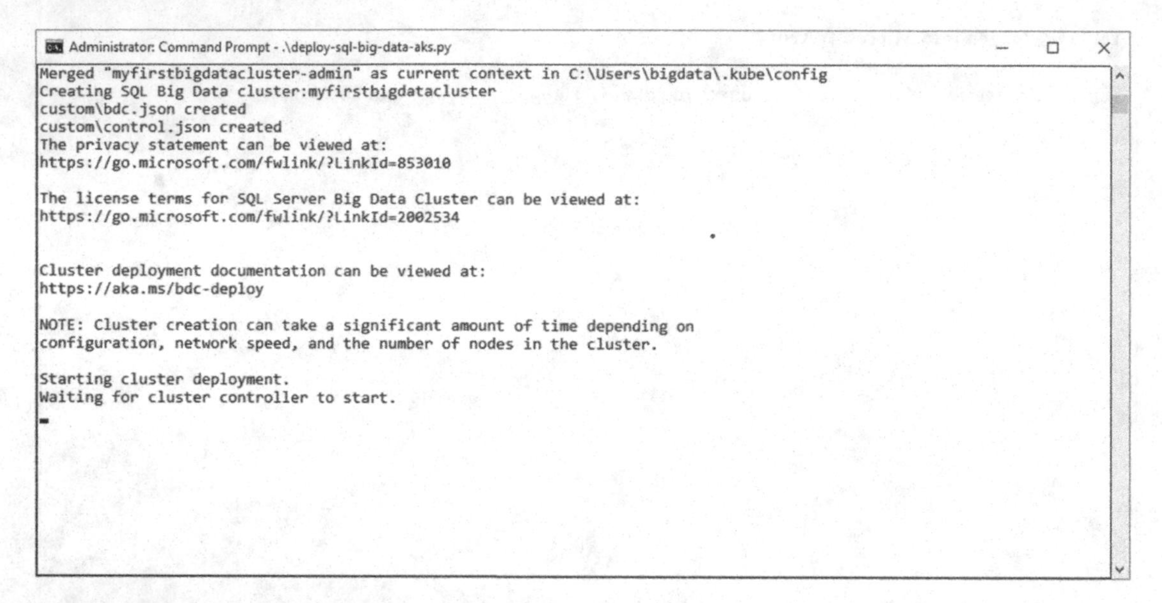


***Figure 3-24.*** *Output of Python deployment script*

The script will also use *azdata bdc config* to create your JSON file.

As this SQL Server 2019 Big Data Cluster is being deployed to Azure, unlike during your local install which you could just reach it using the localhost address, you will need information about the IP addresses and ports of the installation. Therefore, IP addresses and ports are provided at the end as shown in Figure 3-25.

```
Administrator: Command Prompt
Waiting for cluster controller to start.
Waiting for cluster controller to start.
Waiting for cluster controller to start.
Waiting for cluster controller to start.
Waiting for cluster controller to start.
Cluster controller endpoint is available at [illegible].
Cluster control plane is ready.
Data pool is ready.
Compute pool is ready.
Master pool is ready.
Storage pool is ready.
Cluster deployed successfully.
Logged in successfully to `https://137.135.52.89:30080`

SQL Server big data cluster endpoints:
Description                                          Endpoint                                                 Name               Protocol
---------------------------------------------------  -------------------------------------------------------  -----------------  ---------
Gateway to access HDFS files, Spark                  https://[illegible]                                      gateway            https
Spark Jobs Management and Monitoring Dashboard       https://[illegible]/gateway/default/sparkhistory         spark-history      https
Spark Diagnostics and Monitoring Dashboard           https://[illegible]/gateway/default/yarn                 yarn-ui            https
Application Proxy                                    https://[illegible]                                      app-proxy          https
Management Proxy                                     https://[illegible]                                      mgmtproxy          https
Log Search Dashboard                                 https://[illegible]/kibana                               logsui             https
Metrics Dashboard                                    https://[illegible]/grafana                              metricsui          https
Cluster Management Service                           https://[illegible]                                      controller         https
SQL Server Master Instance Front-End                 [illegible]                                              sql-server-master  tds
HDFS File System Proxy                               https://[illegible]/gateway/default/webhdfs/v1           webhdfs            https
Proxy for running Spark statements, jobs, applications https://[illegible]/gateway/default/livy/v1            livy               https

C:\Users\bigdata\Desktop>_
```

***Figure 3-25.*** *Final output of Python deployment script including IP addresses*

If you ever forget what your IPs were, you can run this simple script as shown in Listing 3-11.

***Listing 3-11.*** Retrieve Kubernetes service IPs using kubectl

```
kubectl get service -n <clustername>
```

And if you forgot the name of your cluster as well, try Listing 3-12.

***Listing 3-12.*** Retrieve Kubernetes namespaces using kubectl

```
kubectl get namespaces
```

If you are running more than one cluster at a time, the script in Listing 3-13 might also become helpful. Just save it as IP.py and you can run it as shown in Figure 3-26.

***Listing 3-13.*** Python script to retrieve endpoints of a Big Data Cluster

```
CLUSTER_NAME="myfirstbigdatacluster"
from subprocess import check_output, CalledProcessError, STDOUT, Popen,
PIPE
import os
import getpass
def executeCmd (cmd):
    if os.name=="nt":
        process = Popen(cmd.split(),stdin=PIPE, shell=True)
    else:
        process = Popen(cmd.split(),stdin=PIPE)
    stdout, stderr = process.communicate()
    if (stderr is not None):
        raise Exception(stderr)
print("")
print("SQL Server big data cluster connection endpoints: ")
print("SQL Server master instance:")
command="kubectl get service master-svc-external -o=custom-columns=""IP:.
status.loadBalancer.ingress[0].ip,PORT:.spec.ports[0].port"" -n "+CLUSTER_
NAME
executeCmd(command)
print("")
```

```
print("HDFS/KNOX:")
command="kubectl get service gateway-svc-external -o=custom-
columns=""IP:status.loadBalancer.ingress[0].ip,PORT:.spec.ports[0].port""
-n "+CLUSTER_NAME
executeCmd(command)
print("")
print("Cluster administration portal (https://<ip>:<port>):")
command="kubectl get service mgmtproxy-svc-external -o=custom-
columns=""IP:status.loadBalancer.ingress[0].ip,PORT:.spec.ports[0].port""
-n "+CLUSTER_NAME
executeCmd(command)
print("")
```

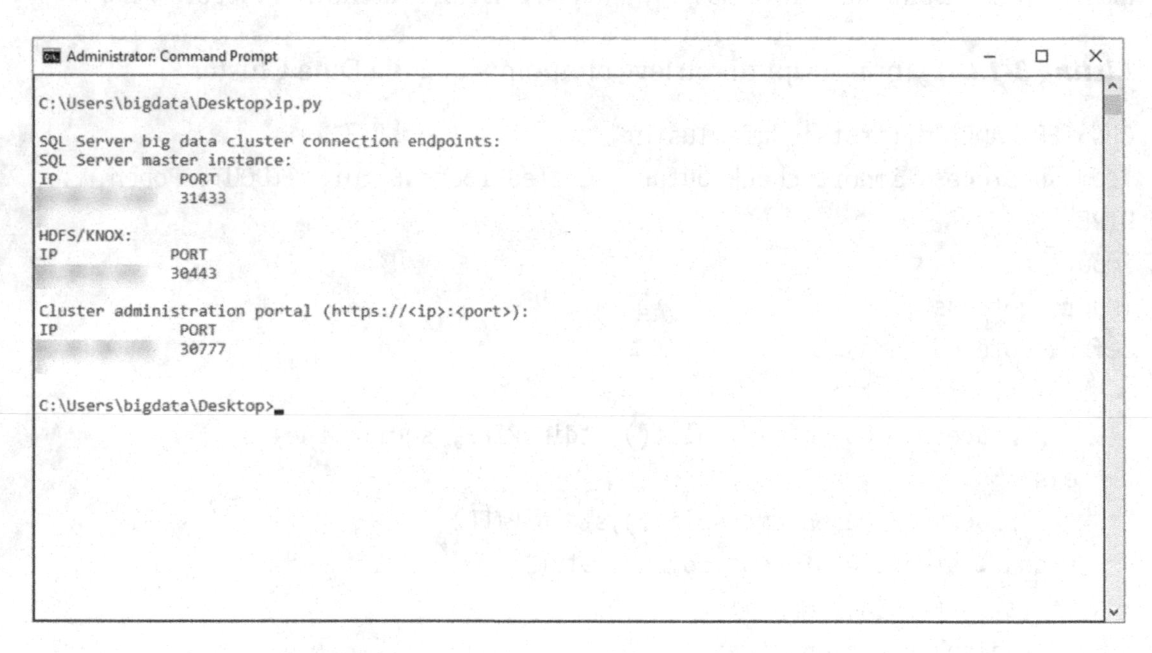


***Figure 3-26.*** *Output of IP.py*

You're done! Your Big Data Cluster in Azure Kubernetes Service is now up and running.

---

**Note** Whether you use it or not, this cluster will accumulate cost based on the number of VMs and their size so it's a good idea not to leave it idling around!

---

## Deploy Your Big Data Cluster Through Azure Data Studio

If you prefer a graphical wizard for your deployment, the answer is Azure Data Studio (ADS)! ADS provides you multiple options to deploy SQL Server, and Big Data Clusters are among them. In ADS, locate the link "New Deployment" which can be found on the welcome screen as well as in a context menu next to your active connections as shown in Figure 3-27.

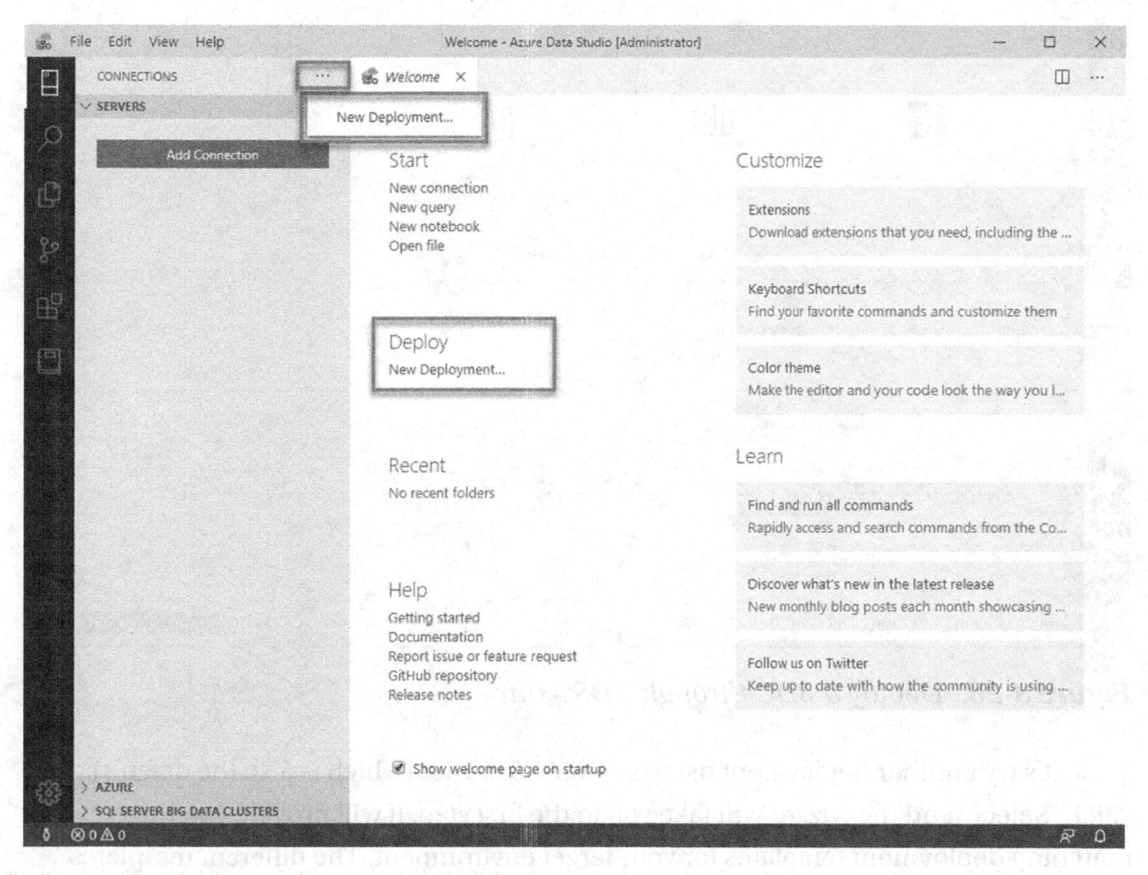


***Figure 3-27.*** *New Deployment in ADS*

On the following screen, select "SQL Server Big Data Cluster." The wizard will ask you to accept the license terms, select a version, and also pick a deployment target. Supported targets for this wizard are currently a new Azure Kubernetes Service (AKS) cluster, an existing AKS cluster, or an existing kubeadm cluster. If you plan to deploy toward an existing cluster, the Kubernetes contexts/connections need to be present in your Kubernetes configuration. If the Kubernetes cluster was not created from the same machine, it's probably still missing. In this case, you can either copy the *.kube* file to your local machine or configure Kubernetes manually as described at `https://kubernetes.io/docs/tasks/access-application-cluster/access-cluster/`.

On the lower end of the screen, the wizard will also list the required tools again and confirm whether all of them are installed in the appropriate version as shown in Figure 3-28.

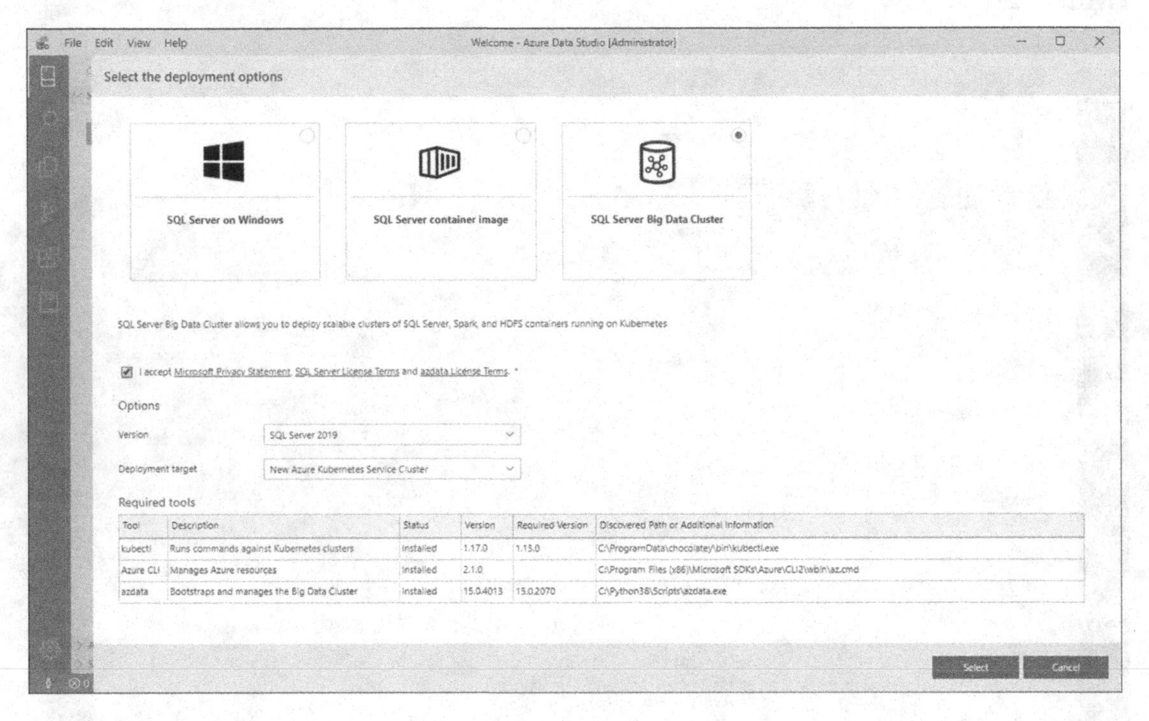


***Figure 3-28.*** *Deploy a BDC through ADS – intro*

Let's try another deployment using a new AKS cluster (which is also the default). Click "Select" and the wizard will take you to the first step. It will provide you the matching deployment templates for your target environment. The different templates differ by size as well as features, like authentication type and high availability, as you can see in Figure 3-29.

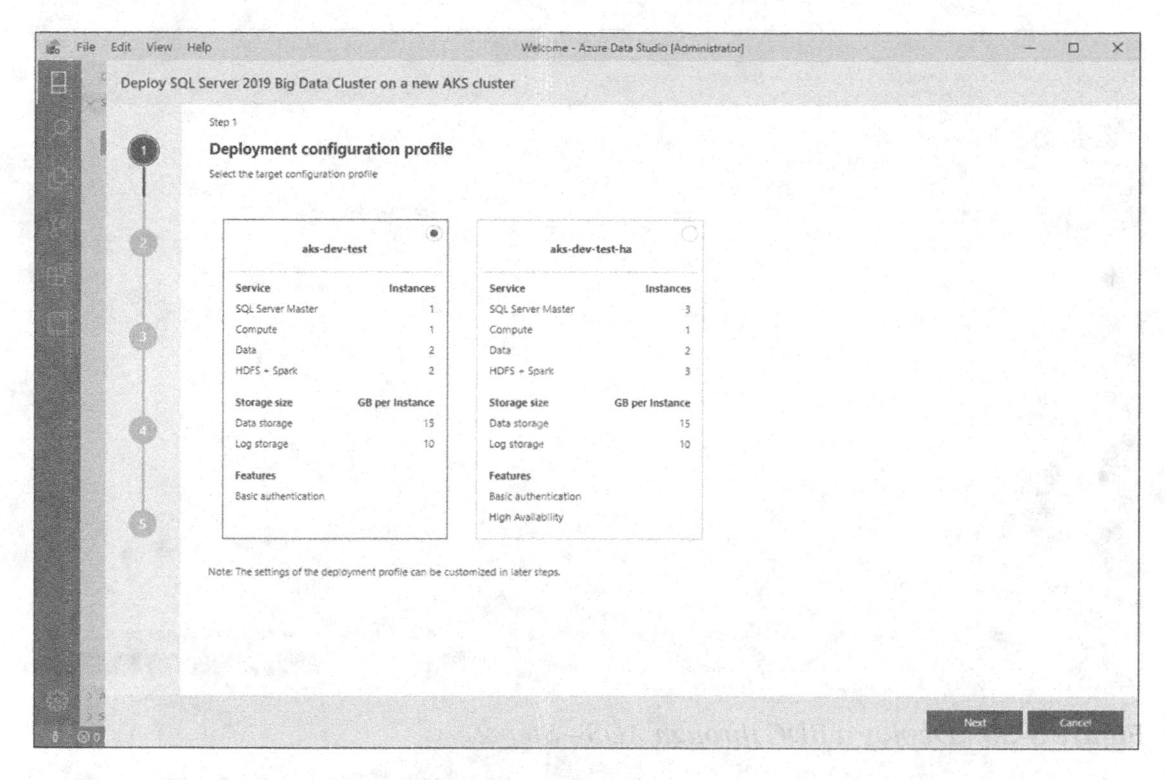


***Figure 3-29.** Deploy a BDC through ADS - Step 1*

The following screen will depend on your target. As we chose to deploy to Azure including a new cluster, we need to provide a subscription, resource group name, location, cluster name, as well as the number and size of the underlying VMs (see Figure 3-30).

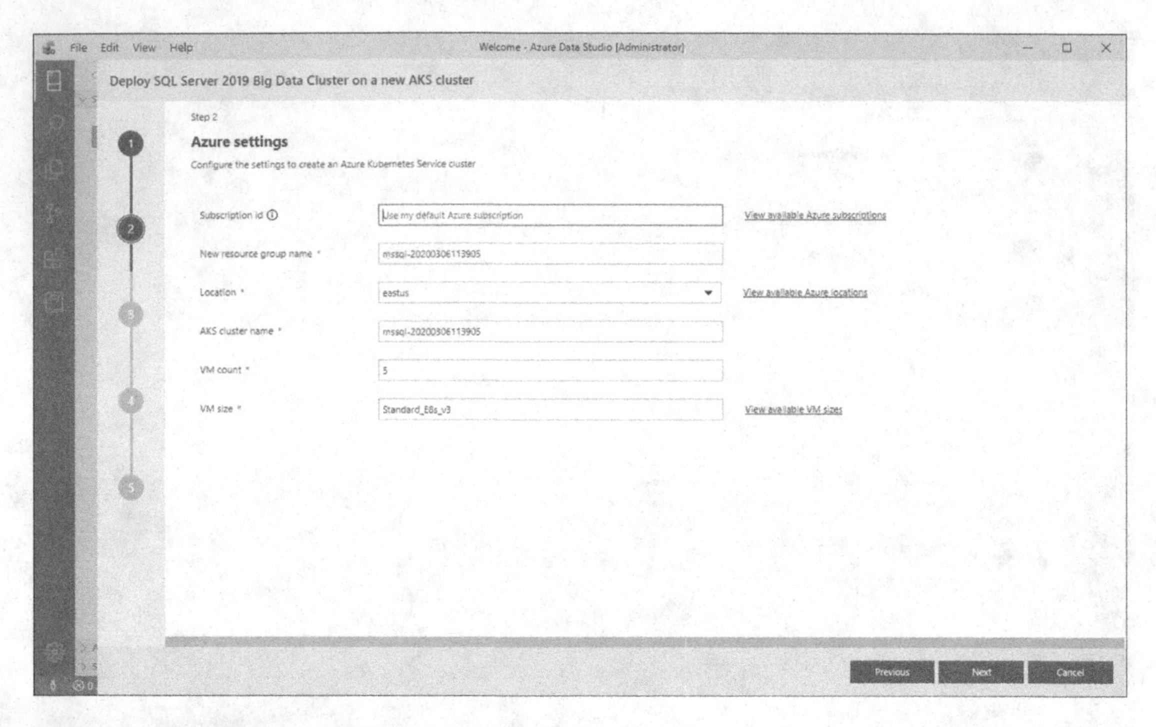


***Figure 3-30.*** *Deploy a BDC through ADS – Step 2*

In Step 3, as illustrated in Figure 3-31, we define the name of the Big Data Cluster (unlike in the previous step where we've set the name for the Kubernetes cluster!) as well as the authentication type.

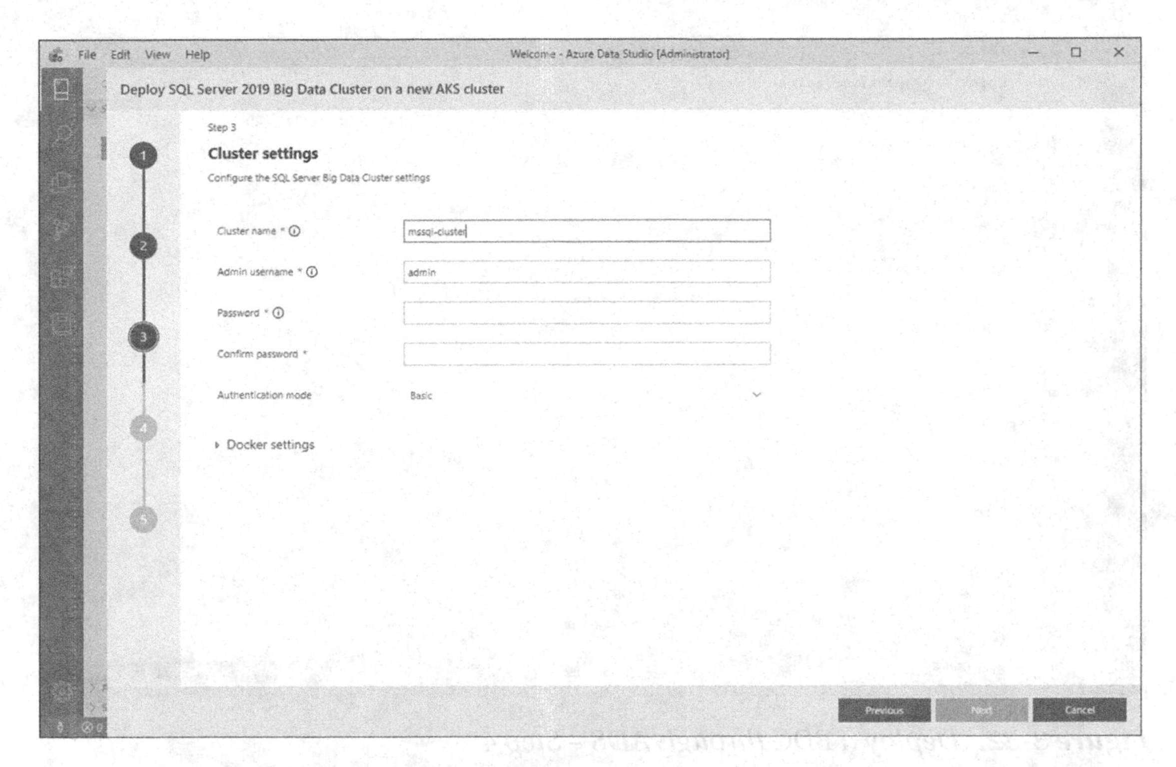


***Figure 3-31.** Deploy a BDC through ADS – Step 3*

In the last configuration screen which we show in Figure 3-32, you can modify the number of instances per pool as well as claim sizes and storage classes for data and logs.

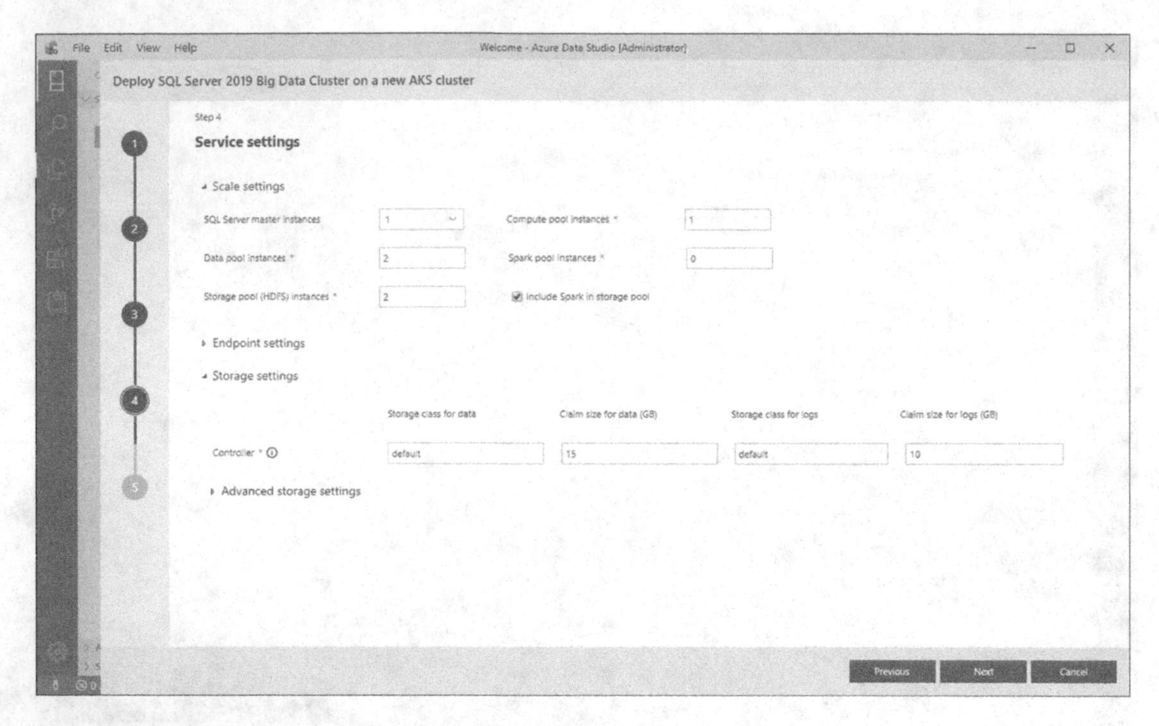


***Figure 3-32.*** *Deploy a BDC through ADS – Step 4*

The final screen as shown in Figure 3-33 gives you a summary of your configuration. If you want to proceed, hit "Script to Notebook"; otherwise, you can navigate back using the "Previous" button to make any necessary changes and adjustments.

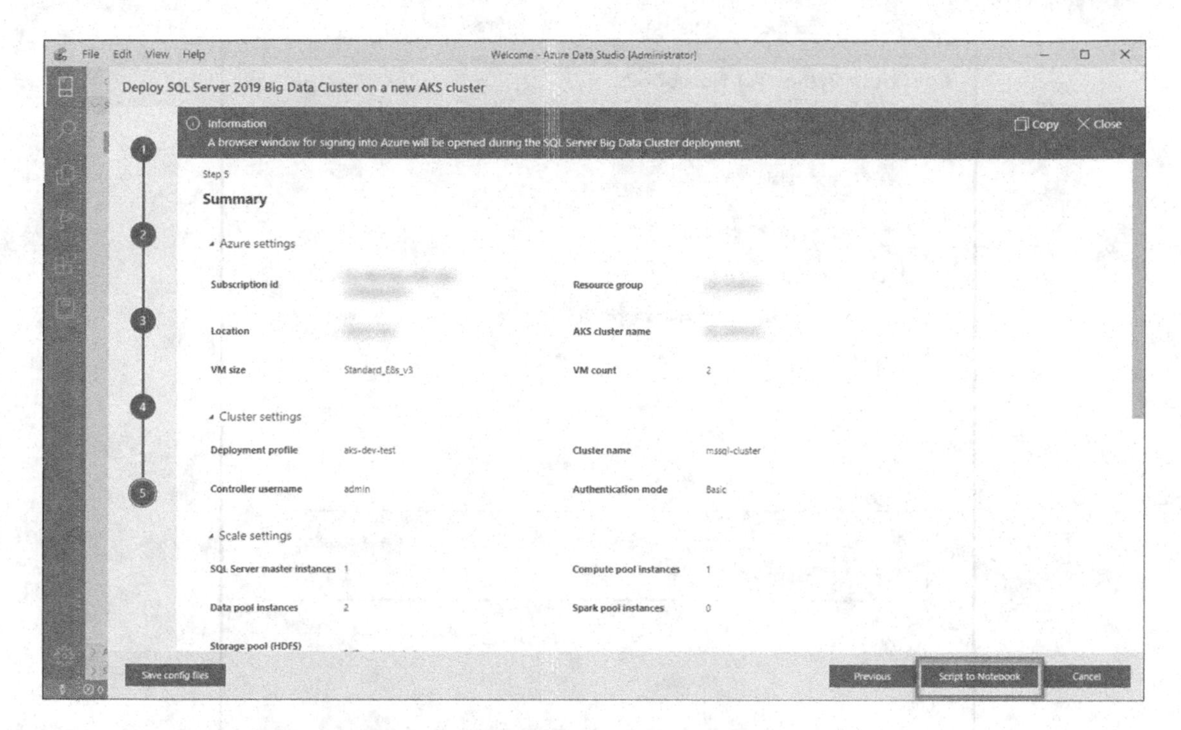


***Figure 3-33.*** *Deploy a BDC through ADS – Summary*

Unless you did so before, ADS will prompt you to install Python for notebooks as shown in Figure 3-34.

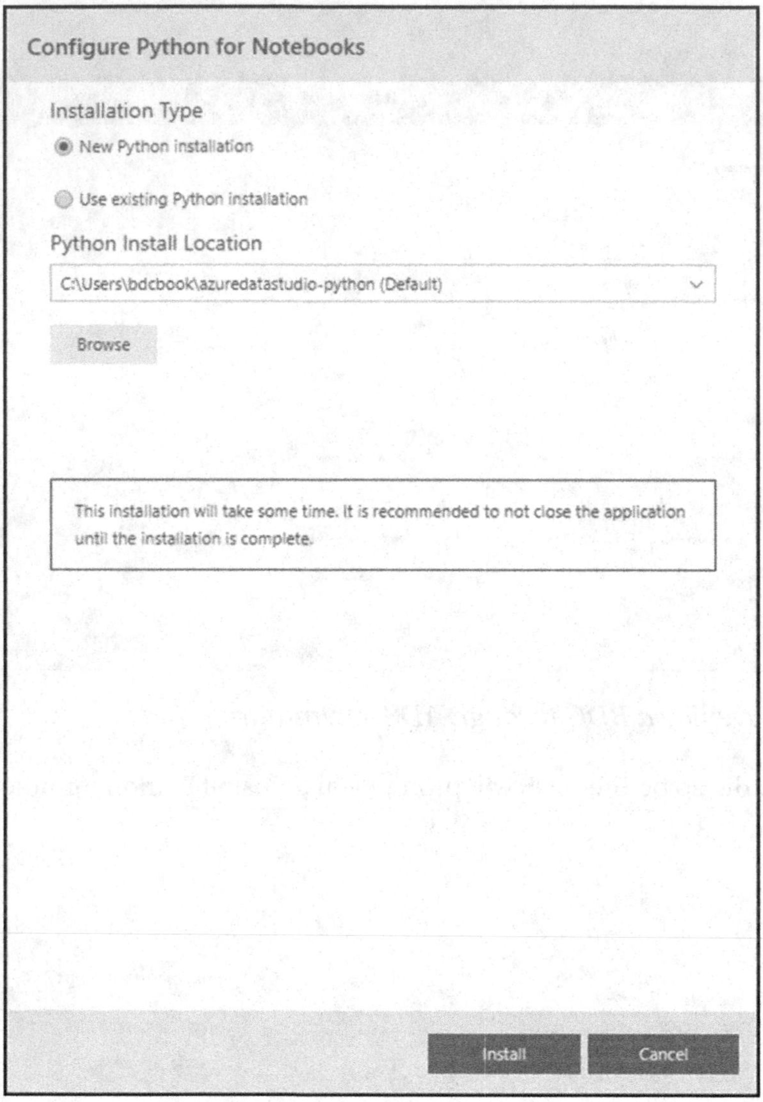

*Figure 3-34. Deploy a BDC through ADS - install Python*

Wait for the installation to complete. All your settings have been populated to a Python notebook which you could either save and store for later or run right away. To execute the notebook, simply click "Run Cells" as shown in Figure 3-35. Just make sure that the Python installation has finished. The kernel combo box should read "Python 3". If it's still showing "Loading kernels...", be patient 😊.

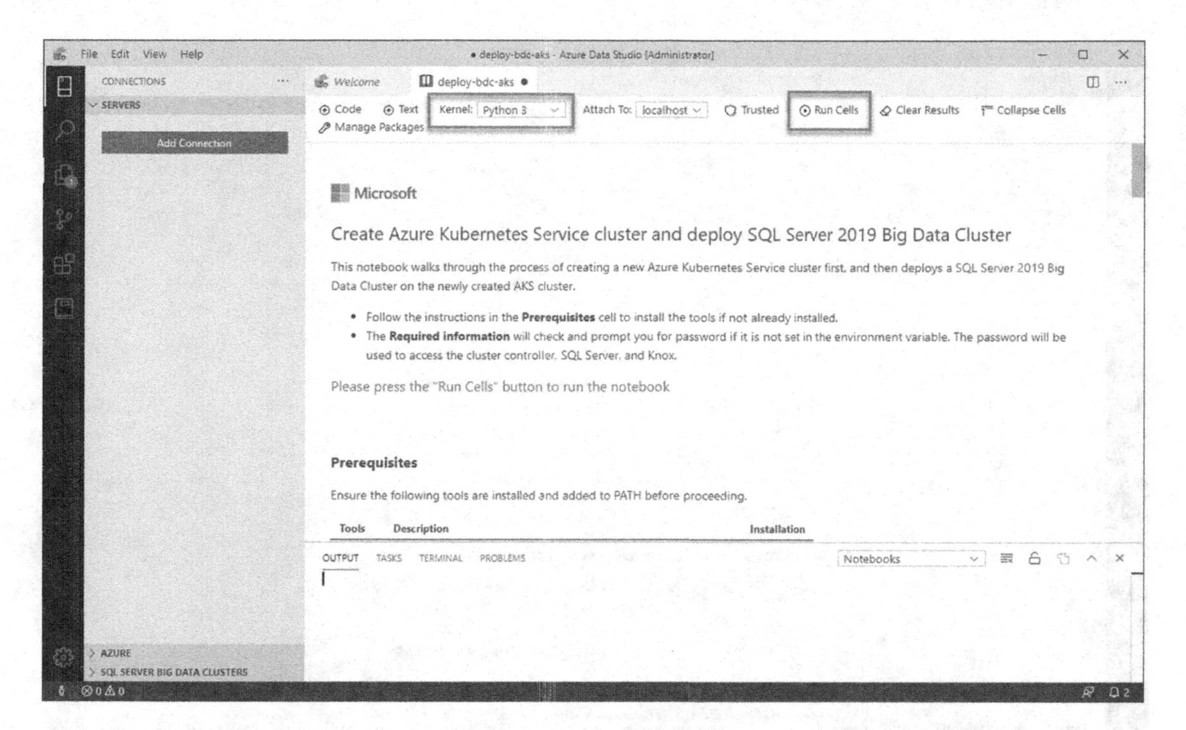


***Figure 3-35.** Deploy a BDC through ADS – notebook predeployment*

Once you click "Run Cells," the deployment process will run through and – unless there are any problems on the way – will report back with the cluster's endpoints at the end, as you can see in Figure 3-36. You will also get a direct link to connect to the master instance. The deployment will take as long as it would with the same parameters using the scripted deployment option.

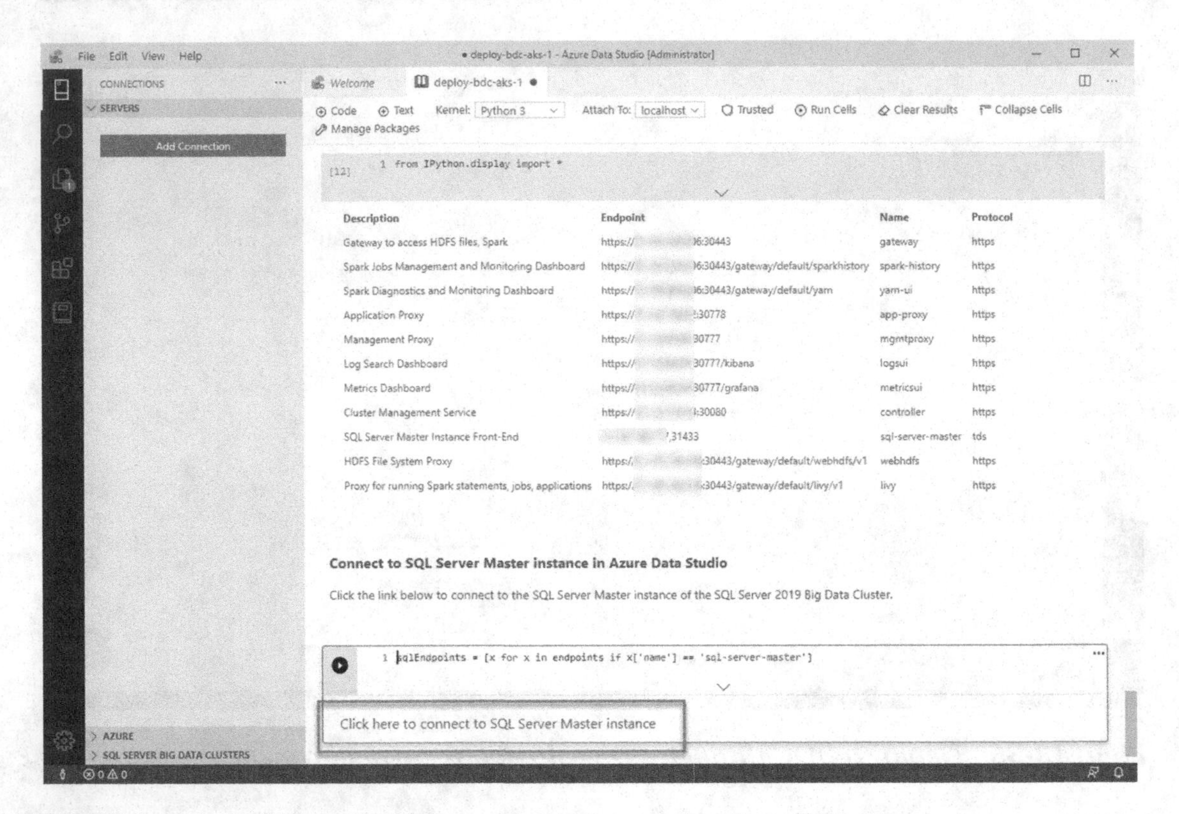


***Figure 3-36.** Deploy a BDC through ADS – notebook postdeployment*

# What Is azdata?

As mentioned before, no matter which path of deployment you choose, the deployment of your Big Data Cluster will always be controlled through a tool call azdata. It's a command-line tool that will help you to create a Big Data Cluster configuration, deploy your Big Data Cluster, and later potentially delete or upgrade your existing cluster.

The logical first step (which is somehow happening behind the scenes in the previous scripts) is to create a configuration as shown in Listing 3-14.

***Listing 3-14.*** Create cluster config using azdata

```
azdata bdc config init [--target -t] [--src -s]
```

Target is just the folder name for your config files (*bdc.json* and *control.json*). The src is one of the existing base templates to start with.

Possible values are (at the time of writing)

- aks-dev-test
- aks-dev-test-ha
- kubeadm-dev-test
- kubeadm-prod

These match the options that you saw when deploying in Azure Data Studio. You can always get all valid options by running *azdata bdc config init -t <yourtarget>* without specifying a source. Keep in mind that these are just templates. If your preferred environment is not offered as a choice, this doesn't necessarily imply that it's not supported, just that you will need to make some adjustments to an existing template to make it match your target. The output is shown is Figure 3-37.

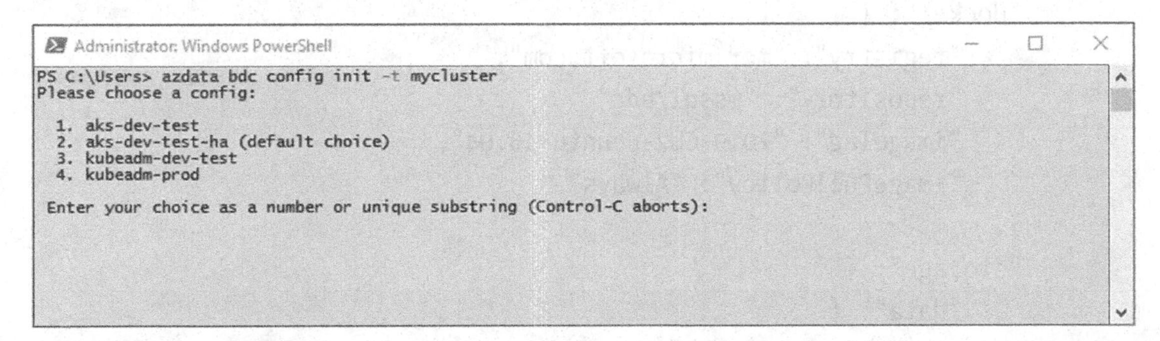


***Figure 3-37.*** *Output of azdata bdc config init without specifying a source*

The source to choose will depend on your deployment type.

Running *azdata bdc config init* results in two .JSON files - bdc.json and control.json - to be created in a subfolder which is named after your target. This will be based on defaults, so we need to make some changes to the configuration. This can be done either using any text editor or using *azdata* again using the *config replace* option, as shown in Listing 3-15, where we use it to modify the name of the Big Data Cluster in the bdc.json file.

***Listing 3-15.*** Modify cluster config using azdata

```
azdata bdc config replace -c myfirstbigdatacluster/bdc.json -j metadata.
name=myfirstbigdatacluster
```

The *control* file defines more general settings like which version, repository, and so on you want to use, whereas the *bdc* file configures the actual setup of your Big Data Cluster environment like the number of replicas per role and so on, as shown in Listings 3-16 and 3-17.

***Listing 3-16.*** Sample control.json

```
{
    "apiVersion": "v1",
    "metadata": {
        "kind": "Cluster",
        "name": "mssql-cluster"
    },
    "spec": {
        "docker": {
            "registry": "mcr.microsoft.com",
            "repository": "mssql/bdc",
            "imageTag": "2019-CU2-ubuntu-16.04",
            "imagePullPolicy": "Always"
        },
        "storage": {
            "data": {
                "className": "",
                "accessMode": "ReadWriteOnce",
                "size": "15Gi"
            },
            "logs": {
                "className": "",
                "accessMode": "ReadWriteOnce",
                "size": "10Gi"
            }
        },
        "endpoints": [
            {
                "name": "Controller",
                "dnsName": "",
```

```
                "serviceType": "NodePort",
                "port": 30080
            },
            {
                "name": "ServiceProxy",
                "dnsName": "",
                "serviceType": "NodePort",
                "port": 30777
            }
        ],
        "settings": {
            "controller": {
                "logs.rotation.size": "5000",
                "logs.rotation.days": "7"
            }
        }
    },
    "security": {
        "activeDirectory": {
            "ouDistinguishedName": "",
            "dnsIpAddresses": [],
            "domainControllerFullyQualifiedDns": [],
            "domainDnsName": "",
            "clusterAdmins": [],
            "clusterUsers": []
        }
    }
}
```

***Listing 3-17.*** Sample bdc.json

```
{
    "apiVersion": "v1",
    "metadata": {
        "kind": "BigDataCluster",
        "name": "mssql-cluster"
    },
    "spec": {
        "resources": {
            "nmnode-0": {
                "spec": {
                    "replicas": 2
                }
            },
            "sparkhead": {
                "spec": {
                    "replicas": 2
                }
            },
            "zookeeper": {
                "spec": {
                    "replicas": 3
                }
            },
            "gateway": {
                "spec": {
                    "replicas": 1,
                    "endpoints": [
                        {
                            "name": "Knox",
                            "dnsName": "",
                            "serviceType": "NodePort",
                            "port": 30443
                        }
                    ]
                }
            },
```

```
"appproxy": {
    "spec": {
        "replicas": 1,
        "endpoints": [
            {
                "name": "AppServiceProxy",
                "dnsName": "",
                "serviceType": "NodePort",
                "port": 30778
            }
        ]
    }
},
"master": {
    "metadata": {
        "kind": "Pool",
        "name": "default"
    },
    "spec": {
        "type": "Master",
        "replicas": 3,
        "endpoints": [
            {
                "name": "Master",
                "dnsName": "",
                "serviceType": "NodePort",
                "port": 31433
            },
            {
                "name": "MasterSecondary",
                "dnsName": "",
                "serviceType": "NodePort",
                "port": 31436
            }
        ],
```

```
                "settings": {
                    "sql": {
                        "hadr.enabled": "true"
                    }
                }
            }
        },
        "compute-0": {
            "metadata": {
                "kind": "Pool",
                "name": "default"
            },
            "spec": {
                "type": "Compute",
                "replicas": 1
            }
        },
        "data-0": {
            "metadata": {
                "kind": "Pool",
                "name": "default"
            },
            "spec": {
                "type": "Data",
                "replicas": 2
            }
        },
        "storage-0": {
            "metadata": {
                "kind": "Pool",
                "name": "default"
            },
            "spec": {
                "type": "Storage",
                "replicas": 3,
```

```
                "settings": {
                    "spark": {
                        "includeSpark": "true"
                    }
                }
            }
        }
    },
    "services": {
        "sql": {
            "resources": [
                "master",
                "compute-0",
                "data-0",
                "storage-0"
            ]
        },
        "hdfs": {
            "resources": [
                "nmnode-0",
                "zookeeper",
                "storage-0",
                "sparkhead"
            ],
            "settings": {
                "hdfs-site.dfs.replication": "3"
            }
        },
        "spark": {
            "resources": [
                "sparkhead",
                "storage-0"
            ],
```

```
                    "settings": {
                        "spark-defaults-conf.spark.driver.memory": "2g",
                        "spark-defaults-conf.spark.driver.cores": "1",
                        "spark-defaults-conf.spark.executor.instances": "3",
                        "spark-defaults-conf.spark.executor.memory": "1536m",
                        "spark-defaults-conf.spark.executor.cores": "1",
                        "yarn-site.yarn.nodemanager.resource.memory-mb": "18432",
                        "yarn-site.yarn.nodemanager.resource.cpu-vcores": "6",
                        "yarn-site.yarn.scheduler.maximum-allocation-mb": "18432",
                        "yarn-site.yarn.scheduler.maximum-allocation-vcores": "6",
                        "yarn-site.yarn.scheduler.capacity.maximum-am-resource-
                        percent": "0.3"
                    }
                }
            }
        }
}
```

As you can see, the file allows you to change quite a lot of settings. While you may leave many of them at their default, this comes in quite handy, especially in terms of storage. You can change the disk sizes as well as the storage type. For more information on storage in Kubernetes, we recommend reading `https://kubernetes.io/docs/concepts/storage/`.

All deployments use persistent storage by default. Unless you have a good reason to change that, you should keep it that way as nonpersistent storage can leave your cluster in a nonfunctioning state in case of restarts, for example.

Run the following command (Listing 3-18) in a command prompt where you've set the environment variables before.

***Listing 3-18.*** Create cluster using azdata

```
azdata bdc create -c myfirstbigdatacluster --accept-eula yes
```

Now sit back, relax, follow the output of azdata as shown in Figure 3-38, and wait for the deployment to finish.

```
Administrator: Command Prompt - azdata bdc create -c myfirstbigdatacluster --accept-eula yes

C:\Users\bigdata\Desktop>azdata bdc create -c myfirstbigdatacluster --accept-eula yes
The privacy statement can be viewed at:
https://go.microsoft.com/fwlink/?LinkId=853010

The license terms for SQL Server Big Data Cluster can be viewed at:
https://go.microsoft.com/fwlink/?LinkId=2002534

Cluster deployment documentation can be viewed at:
https://aka.ms/bdc-deploy


NOTE: Cluster creation can take a significant amount of time depending on
configuration, network speed, and the number of nodes in the cluster.

Starting cluster deployment.
Waiting for cluster controller to start.
Waiting for cluster controller to start.
Waiting for cluster controller to start.
Waiting for cluster controller to start.
Cluster controller endpoint is available at [illegible].
Cluster control plane is ready.
Data pool is ready.
Storage pool is ready.
Compute pool is ready.
```

***Figure 3-38.*** *Output of azdata bdc create*

Depending on the size of your machine, this may take anywhere from minutes to hours.

## Others

There are multiple other Kubernetes environments available – from Raspberry Pi to VMWare. Many but not all of them support SQL Server 2019 Big Data Clusters. The number of supported platforms will grow over time, but there is no complete list of compatible environments. If you are looking at a specific setup, the best and easiest way is to just give it a try!

# Advanced Deployment Options

Besides the configuration options mentioned earlier, we would like to point your attention to two additional opportunities to make more out of your Big Data Cluster: Active Directory authentication and HDFS tiering.

## Active Directory Authentication for Big Data Clusters

If you want to use Active Directory (AD) integration rather than basic authentication, this can be achieved through additional information provided in your *control.json* and *bdc.json* files. While *bdc.json* only requires the nameservers to be set to the domain controller's DNS, *control.json* needs a couple of additional parameters, which are shown in Listing 3-19.

***Listing 3-19.*** AD parameters in control.json

```
"security": {
        "activeDirectory": {
            "ouDistinguishedName": "",
            "dnsIpAddresses": [],
            "domainControllerFullyQualifiedDns": [],
            "domainDnsName": "",
            "clusterAdmins": [],
            "clusterUsers": []
        }
```

At the time of writing, there are quite a few limitations though. For example, AD authentication is only supported on kubeadm, not on AKS deployments, and you can only have one Big Data Cluster per domain. You will also need to set up a few very specific objects in your AD before deploying the Big Data Cluster. Please see the official documentation at `https://docs.microsoft.com/en-us/sql/big-data-cluster/deploy-active-directory?view=sql-server-ver15` for detailed steps on how to enable this.

## HDFS Tiering in Big Data Clusters

Should you already have an existing HDFS stored in either Azure Data Lake Store Gen2 or Amazon S3, you can mount this storage as a subdirectory of your Big Data Cluster's HDFS. This will be achieved through a combination of environment variables, kubectl and azdata command. As the process differs slightly per source type, we refer you to the official documentation which can be found at `https://docs.microsoft.com/en-us/sql/big-data-cluster/hdfs-tiering?view=sql-server-ver15`.

Unlike enabling AD authentication, which happens at deployment, HDFS tiering will be configured on an existing Big Data Cluster.

## Summary

In this chapter, we've installed SQL Server 2019 Big Data Clusters using various methods and to different extents.

Now that we have our Big Data Cluster running and ready for some workload, let's move on to Chapter 4 where we'll show and discuss how the cluster can be queried and used.

CHAPTER 4

# Loading Data into Big Data Clusters

With our first SQL Server Big Data Cluster in place, we should have a look at how we can use it. Therefore, we will start by adding some data to it.

## Getting Azure Data Studio Fully Ready for Your Big Data Clusters

While Azure Data Studio can connect to any Big Data Cluster (and also manage and deploy it) by default, we would recommend you install the Data Virtualization extension which provides you wizards helping with the creation on external (virtual) tables.

To install that extension, first navigate to the Extensions menu in Azure Data Studio as shown in Figure 4-1.




B. Weissman and E. van de Laar, *SQL Server Big Data Clusters*,
https://doi.org/10.1007/978-1-4842-5985-6_4

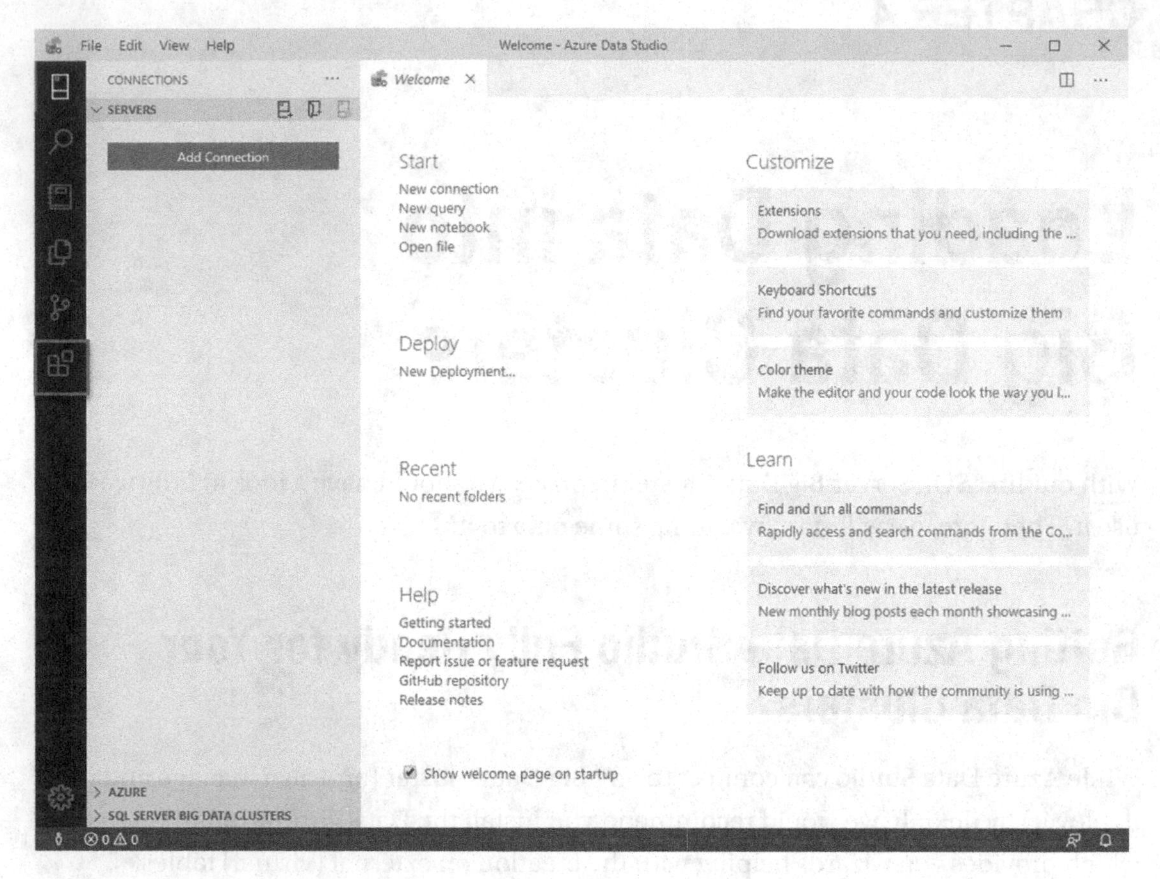


***Figure 4-1.** Install extension from VSIX package in ADS*

Within the extensions marketplace, it should probably already be visible as one of the top recommendations. Otherwise you can also search for it as illustrated in Figure 4-2.

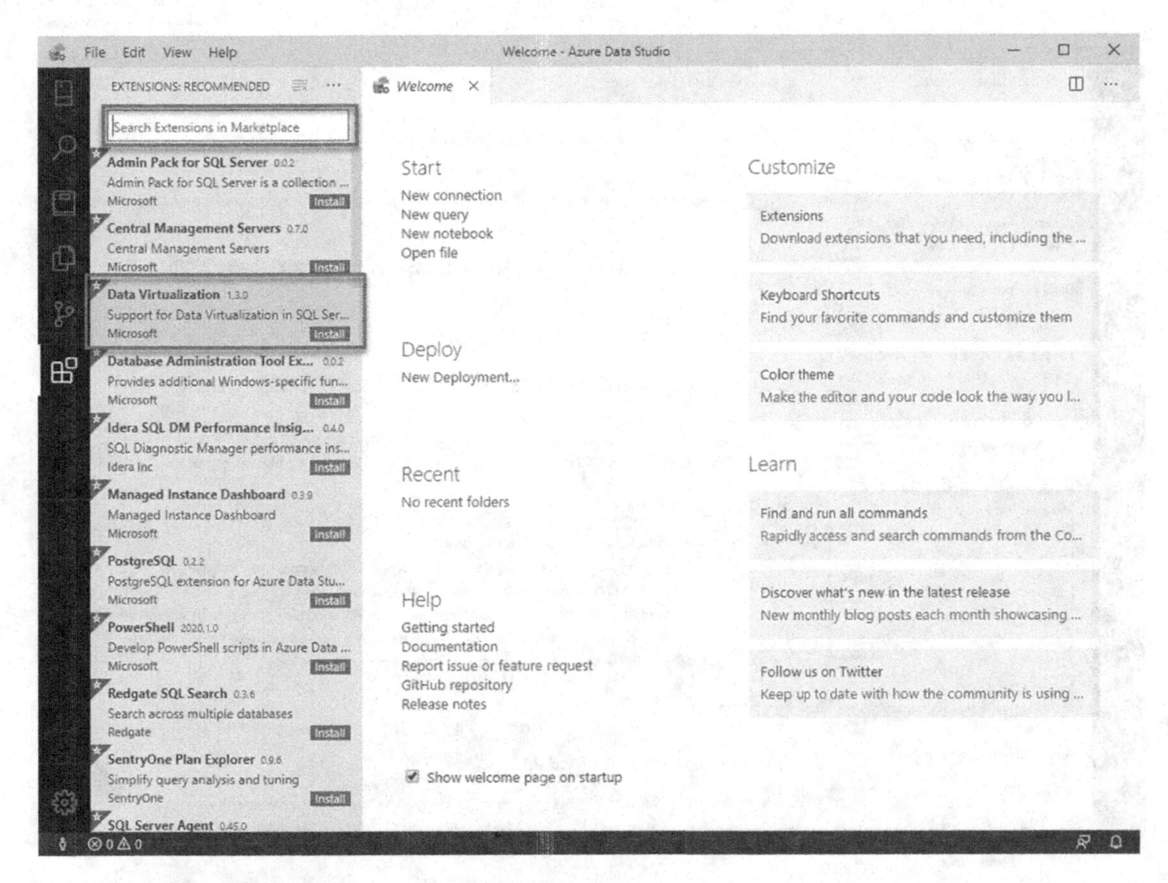


***Figure 4-2.** Extensions in Azure Data Studio*

Click the green "Install" button of the extension and the installation will immediately be triggered as shown in Figure 4-3.

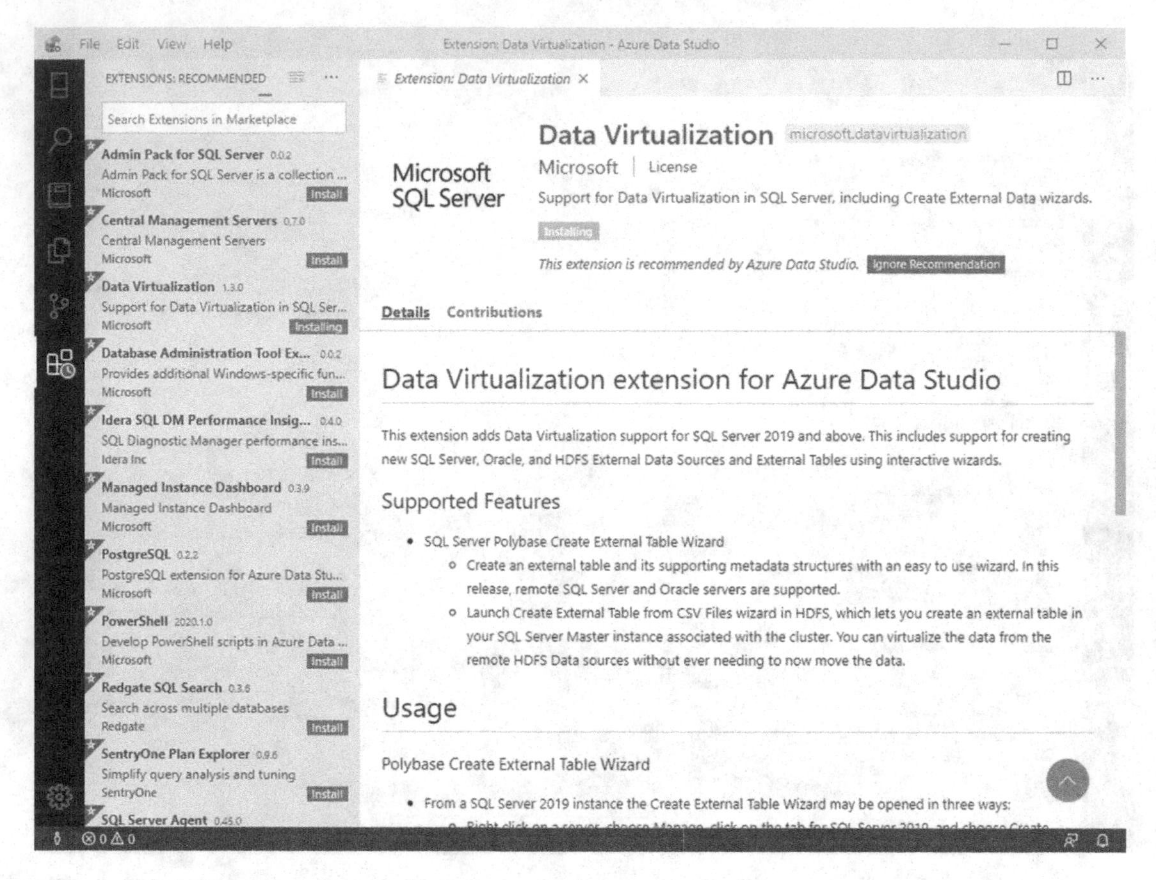


***Figure 4-3.** Extension installation in progress in Azure Data Studio*

The installation usually takes a few minutes and eventually you will see the status of the extension change to "Installed" as shown in Figure 4-4.

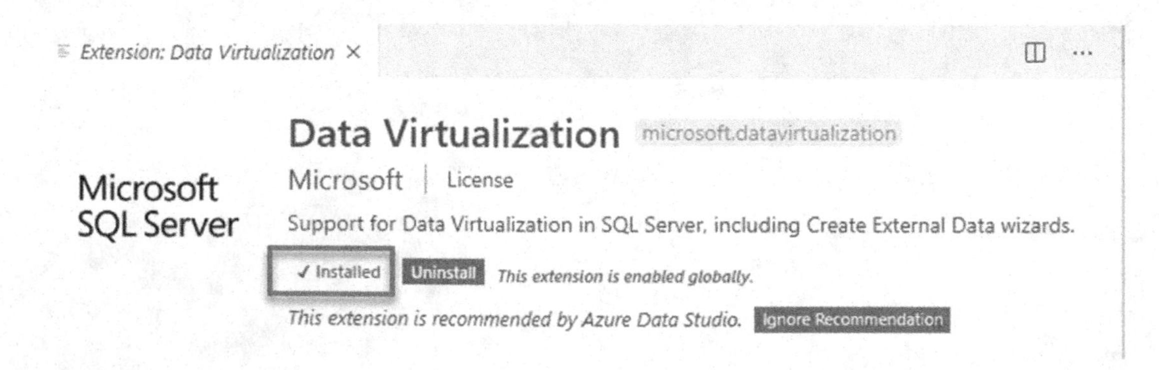


***Figure 4-4.** Installed extension in Azure Data Studio*

The extension is now ready to be used!

# Getting Some Sample Files into the Installation

With everything in place and ready, all we need before we can really get our hands on the new features is some sample data!

## Empty Database

To link some external SQL Server tables into your local instance of SQL Server, the easiest way is to simply create a blank database. Just connect to your SQL Server 2019 instance through either SQL Server Management Studio or Azure Data Studio and create a new database named "BDC_Empty". You can do this through the wizard or by simply running T-SQL as shown in Listing 4-1.

***Listing 4-1.*** Create empty database through T-SQL

```
USE master
GO
CREATE DATABASE BDC_Empty
```

That's it.

## Sample Data Within Your Big Data Cluster

If you went for a full installation including the Kubernetes cluster, there are some easy ways and techniques to push some samples to that. In case you only deployed a local installation with PolyBase enabled but without a Kubernetes cluster, you can skip this part – it wouldn't work anyway.

### Restoring Any SQL Server Backup to Your Master Instance

Assuming an empty database is not enough for you, you may wonder how to restore an existing database to your Master Instance. Let's give that a try with AdventureWorks2014.

If you don't have a backup of AdventureWorks2014 on hand, you can just get it from GitHub, for example, through curl (Listing 4-2).

***Listing 4-2.*** Download AdventureWorks2014 from GitHub using curl

```
curl -L -G "https://github.com/Microsoft/sql-server-samples/releases/
download/adventureworks/AdventureWorks2014.bak" -o AdventureWorks2014.bak
```

Now that we have an actual file to be restored, we need to push that to the Master Instance's filesystem first. This task will be achieved through kubectl (Listing 4-3); you will need to replace your cluster's namespace and master pod name accordingly.

***Listing 4-3.*** Copy AdventureWorks2014 to the Master Instance using kubectl

```
kubectl cp AdventureWorks2014.bak <CLUSTER_NAMESPACE>/<MASTER_POD_
NAME>:var/opt/mssql/data/ -c mssql-server
```

Last but not least, we need to restore the database from the .bak file. This can be achieved through regular T-SQL. In this case, just connect to your master instance and run the script. Of course, for more complex scenarios, you could use sqlcmd with an input file or any other SQL Server mechanism you're comfortable with. Here this includes using the restore wizard in SQL Server Management Studio (Listing 4-4).

***Listing 4-4.*** Restore AdventureWorks2014 to the Master Instance

```
USE [master]
RESTORE DATABASE [AdventureWorks2014] FROM  DISK = N'/var/opt/mssql/
data/AdventureWorks2014.bak' WITH  FILE = 1,  MOVE N'AdventureWorks2014_
Data' TO N'/var/opt/mssql/data/AdventureWorks2014_Data.mdf',  MOVE
N'AdventureWorks2014_Log' TO N'/var/opt/mssql/data/AdventureWorks2014_Log.
ldf',  NOUNLOAD,  STATS = 5
```

## Microsoft Sample Data

We'll start with the sample data provided by Microsoft on their GitHub page, `https://github.com/Microsoft/sql-server-samples/tree/master/samples/features/sql-big-data-cluster`. Download the files "bootstrap-sample-db.sql" and, depending on your operating system, either "bootstrap-sample-db.cmd" (for Windows) or "bootstrap-sample-db.sh" (for Linux).

You can then run the .cmd or .sh file with the following parameters (Listing 4-5).

***Listing 4-5.*** Install default Microsoft samples

```
USAGE: bootstrap-sample-db.cmd <CLUSTER_NAMESPACE> <SQL_MASTER_ENDPOINT>
<KNOX_ENDPOINT> [--install-extra-samples] [SQL_MASTER_PORT] [KNOX_PORT]
To use basic authentication please set AZDATA_USERNAME and AZDATA_PASSWORD
environment variables.
To use integrated authentication, provide the DNS names for the endpoints.
Port can be specified separately if using non-default values.
```

Just pass the information (IPs, password, namespace) you used or were provided during installation of your cluster, and the script will run automatically and pump some sample data to your installation.

The requirements of this script are

- sqlcmd
- bcp
- kubectl
- curl

If you are running this script from the same box that you used for the initial installation, those requirements should already be satisfied.

## Flight Delay Sample Dataset

In addition to the Microsoft samples, let's also add some more external data. A great place to find free datasets is kaggle.com (Figure 4-5).

If you don't have an account with them yet, just sign up for a free account. Otherwise, just log in to your account.

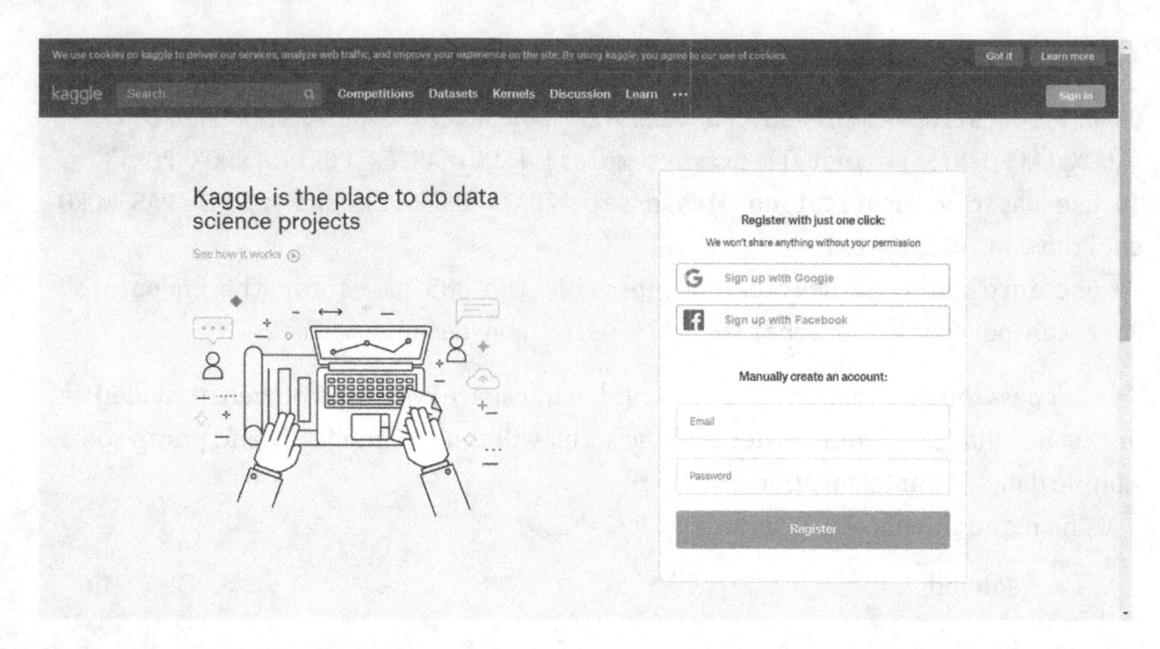


***Figure 4-5.*** *Kaggle.com login*

Once signed in, navigate to Datasets and search for "Flight Delays," which should bring up the "2015 Flight Delays and Cancellations" Dataset from the Department of Transportation as shown in Figure 4-6.

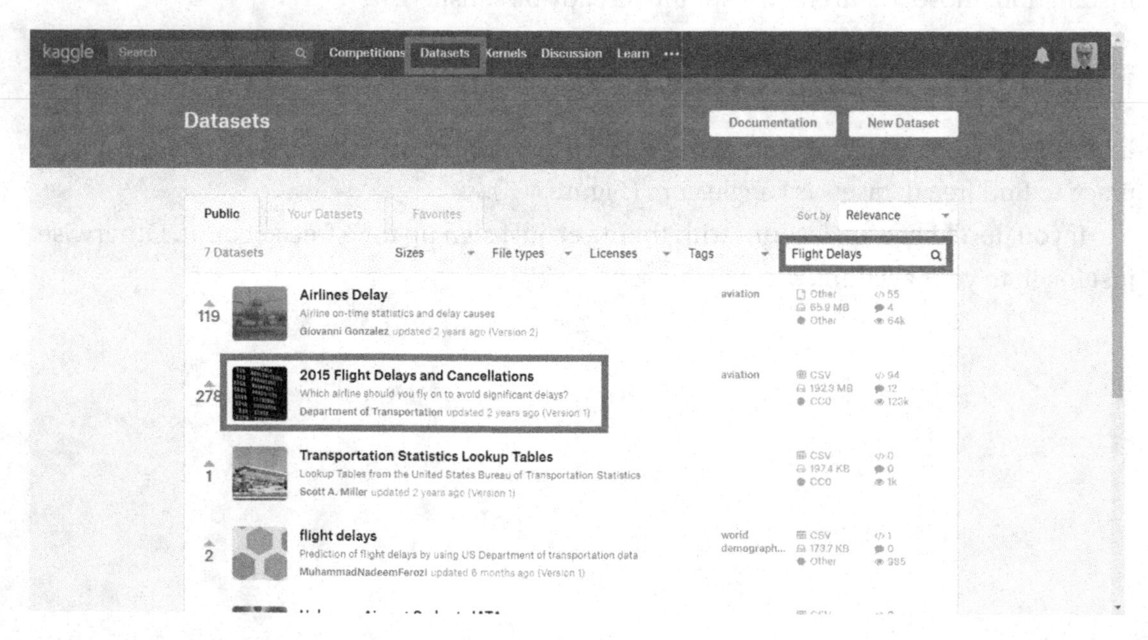


***Figure 4-6.*** *Kaggle.com Datasets*

Alternatively, you can also navigate directly to `www.kaggle.com/usdot/flight-delays`.

The Datasets consist of three files: Airlines, Airports, and Flights. You can download them all at once by clicking "Download," which will trigger one ZIP file containing all files as shown in Figure 4-7.

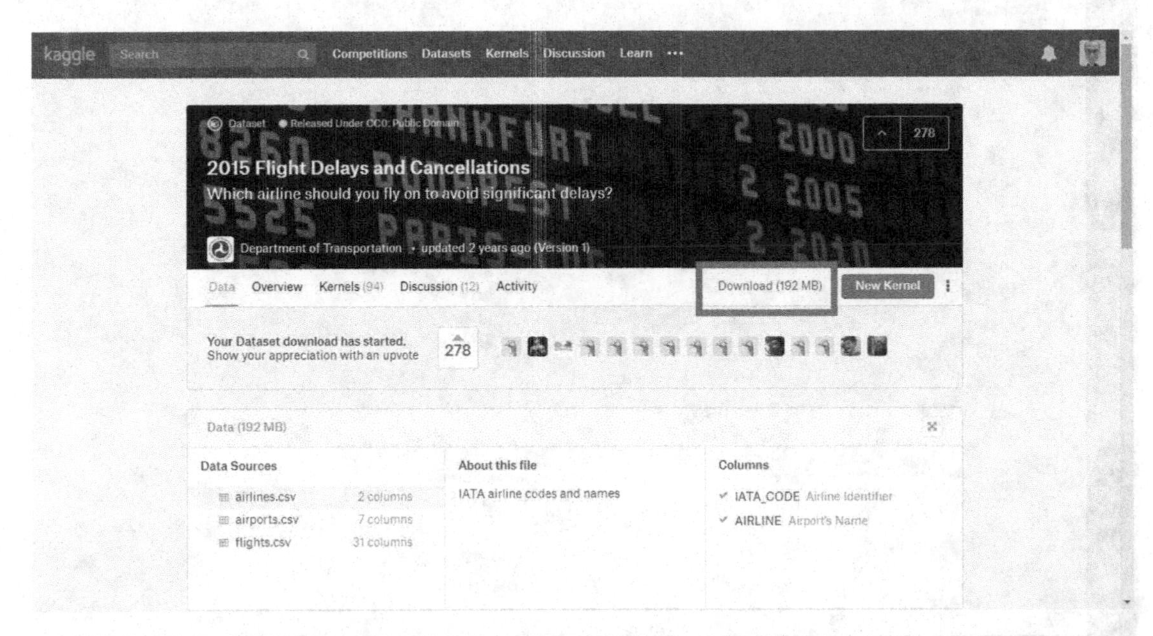


***Figure 4-7.*** *Kaggle.com download Flight Delays Datasets*

While not unreasonably big, this dataset provides a lot of options to explore and work with the data. Once you've downloaded the file, we still need to get that data into our Big Data Cluster. Since these are only three files, we will do this by manually uploading them through Azure Data Studio.

Therefore, connect to your Big Data Cluster in Azure Data Studio, navigate to Data Services, open the HDFS root folder, and create a new directory called "Flight_Delays" as shown in Figure 4-8.

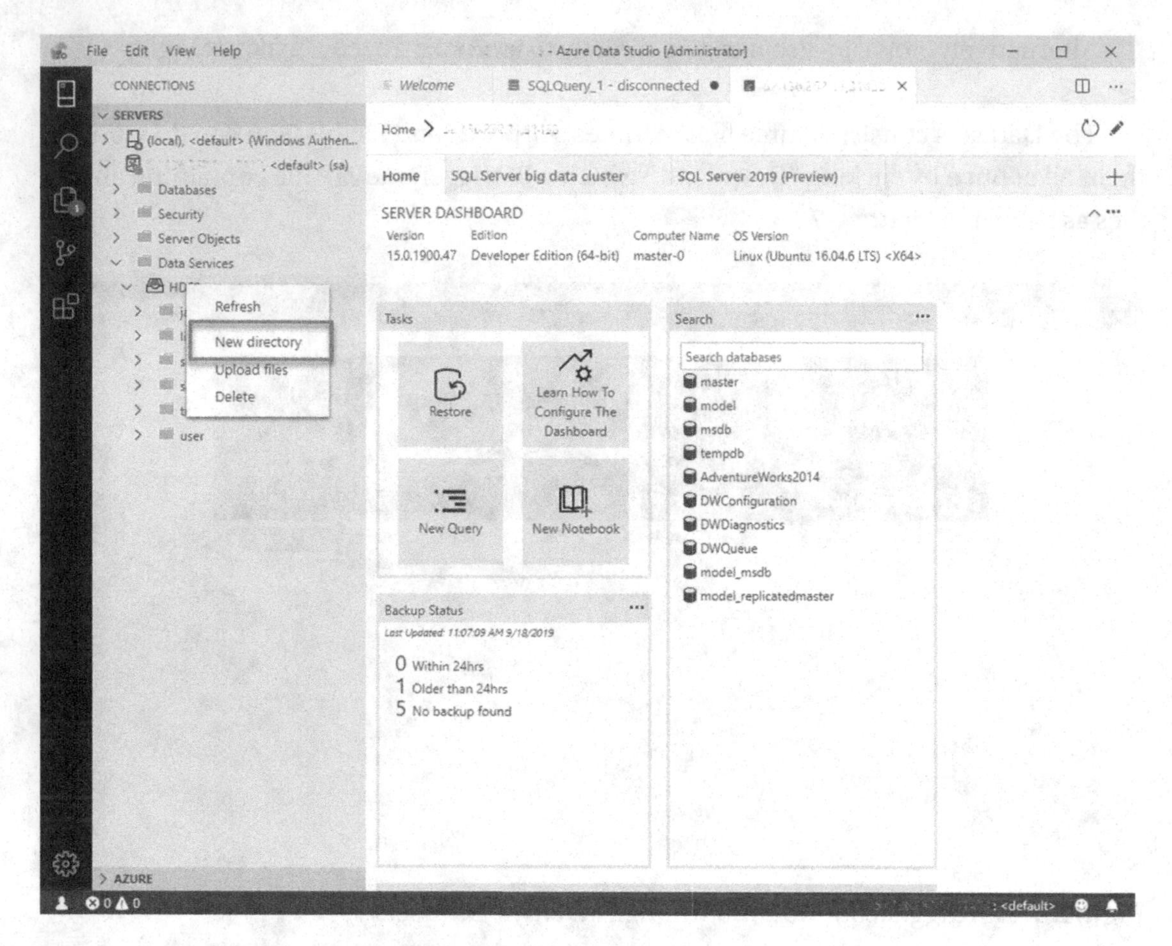


*Figure 4-8. Create new directory on HDFS in ADS*

You can then select this directory, do a right-mouse-click, choose "Upload files," and upload the three CSV files. You can multiselect them so there is no need to upload them one by one. If you do a right-mouse-click and refresh the folder, the files should be visible as shown in Figure 4-9.

*Figure 4-9. Display of files in the new folder in ADS*

The upload progress will also be visible in the footer of Azure Data Studio.

An alternative to the upload through the front end would be to use curl from a command prompt. You can use it both to create the target directory and to upload the actual file.

For just the airlines.csv, this would look like as shown in Listing 4-6 (you would need to replace your IP address and password). The first line will create a directory called "Flight_Delays", while the second line will upload the file "airlines.csv" to it.

***Listing 4-6.*** Upload data to HDFS using curl

```
curl -i -L -k -u root:<yourpassword> -X PUT "https:// <yourIP>/gateway/
default/webhdfs/v1/Flight_Delays?op=MKDIRS"

curl -i -L -k -u root:<yourpassword> -X PUT "https://<yourIP>/gateway/
default/webhdfs/v1/Flight_Delays /airlines.csv?op=create&overwrite=true" -H
"Content-Type: application/octet-stream" -T "airlines.csv"
```

# Azure SQL Database

As described within the use cases in Chapter 1, one way of using the Big Data Cluster PolyBase implementation is to stretch out data to Azure (or any other cloud-based SQL Server for that matter). To get a better feeling of this, unless you already have a database on either another SQL Server or in Azure SQL DB, we recommend to just set up a small database in Azure containing the AdventureWorks Database.

To do so, log on again to the Azure Portal (Figure 4-10) as you did in Chapter 3.

Then pick "Create a resource" on the upper end of your panel on the left and either pick "SQL Database" from the list or search for it.

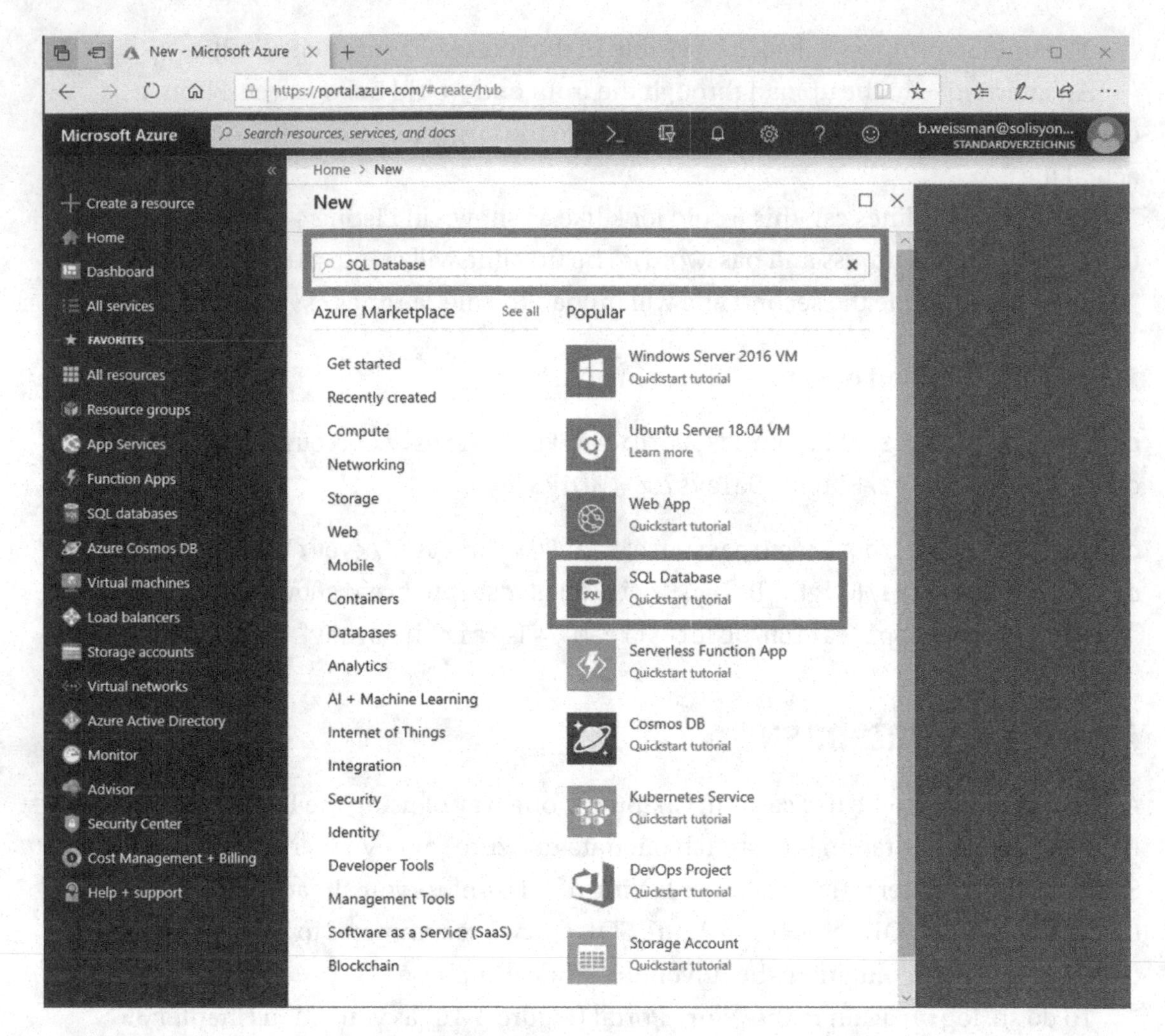


***Figure 4-10.*** *Create resource in Azure Portal*

On the next screen, just click "Create" (Figure 4-11).

*Figure 4-11. Create SQL Database in Azure Portal*

As the name for your database, just use "AdventureWorks," pick the appropriate subscription, create a new resource group or pick an existing, and choose "Sample (AdventureWorksLT)" as your source (Figure 4-12).

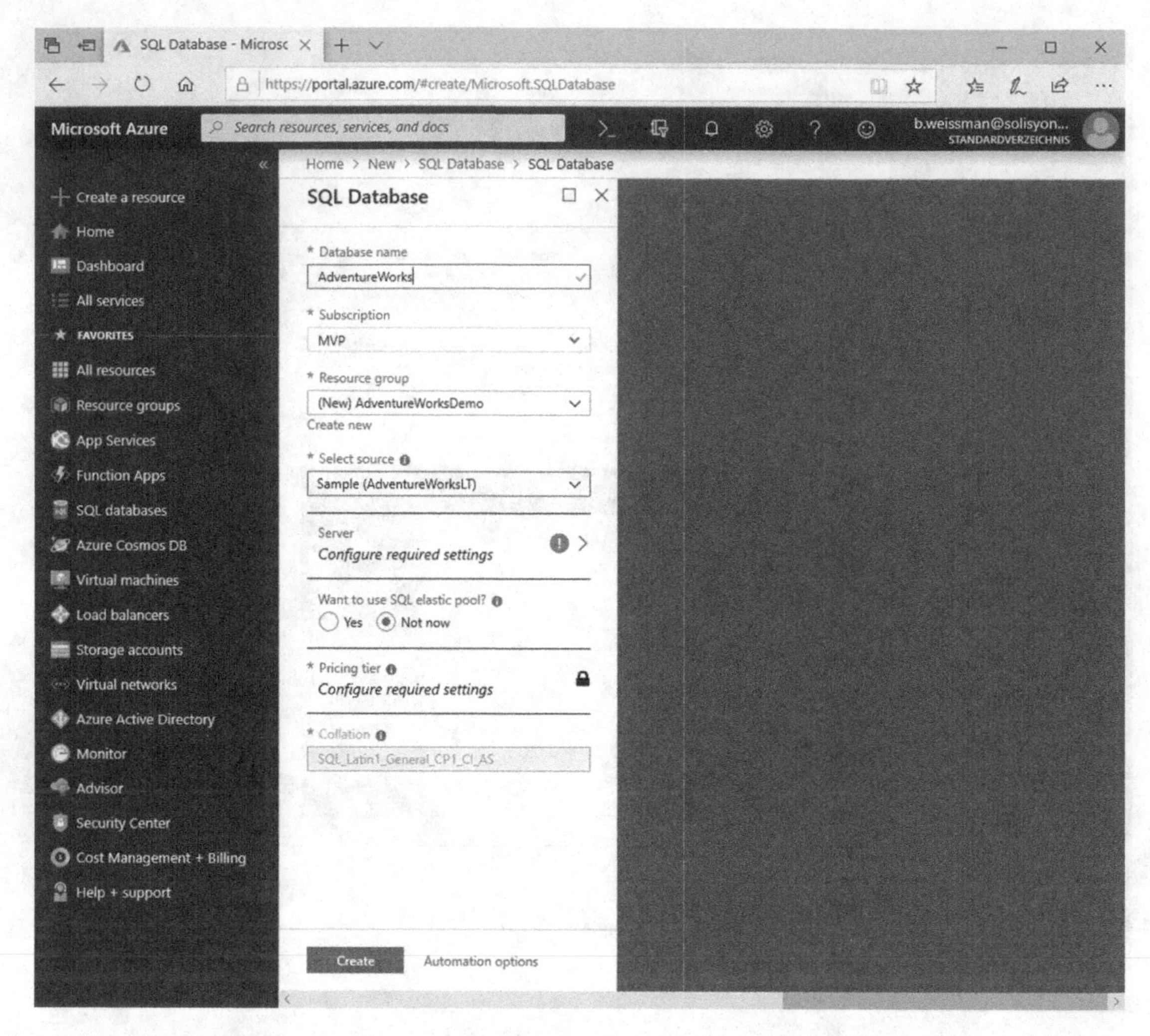


***Figure 4-12.** Configure the SQL Database to be created through Azure Portal*

You will also need to configure a Server (Figure 4-13), therefore expand the Server submenu. Set up the server by providing a name (this has to be unique), a username, and a password. Again, pick the location closest to you and keep the "Allow Azure services to access server" box ticked. This will allow you to access this database from other Azure VMs or services without having to worry about firewall settings. Depending on your setup, you may still need to allow access from your on-premises box – we'll get to that later.

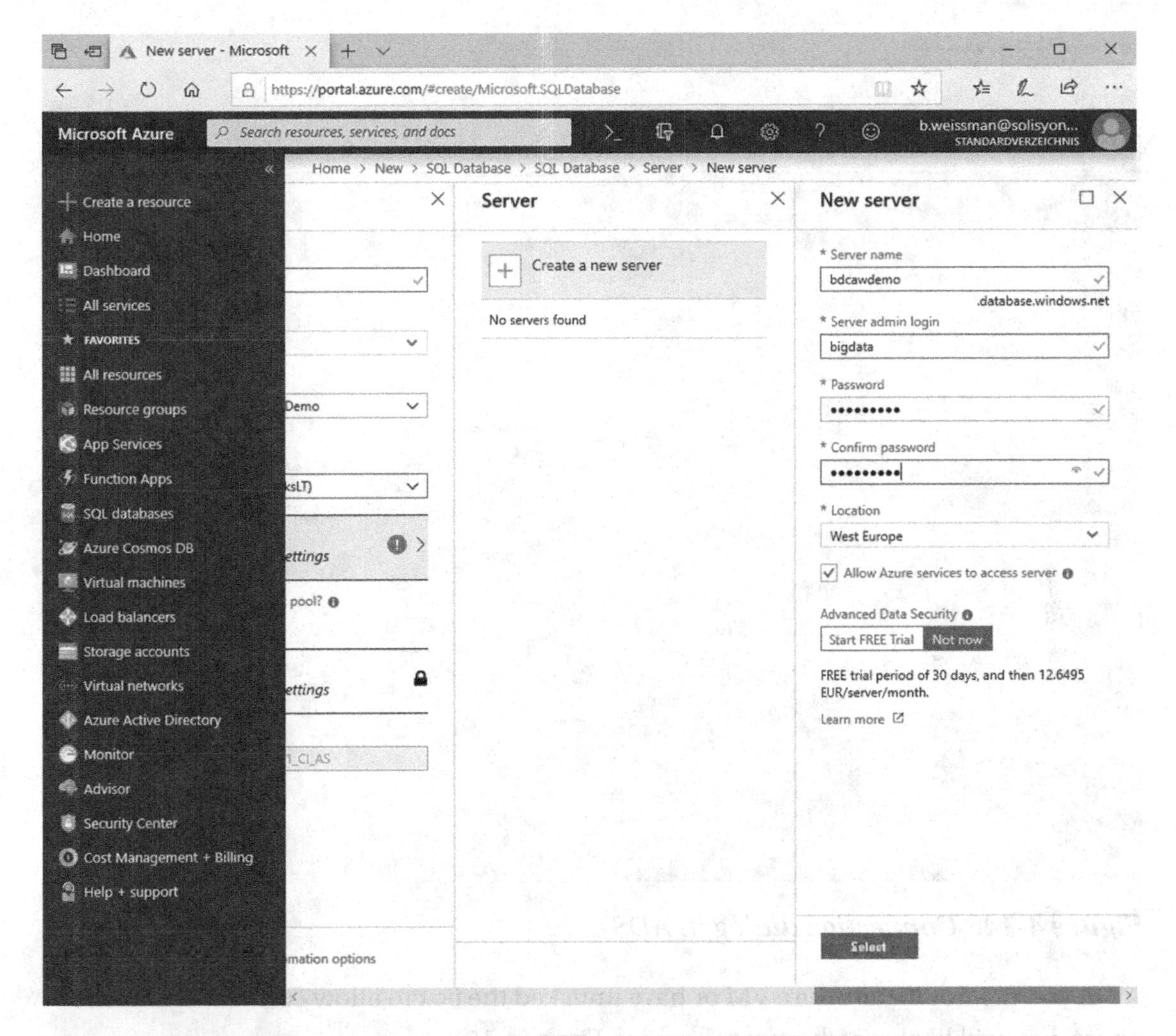


*Figure 4-13. Configure the Server for the new SQL Database in Azure Portal*

You can now change the "Pricing tier" to "Basic" which is the cheapest option but totally sufficient for what we're trying to achieve here.

Confirm your selections by clicking "Create"; this triggers the deployment of the server and the database which should take a couple of minutes. You're done – you have just created the AdventureWorksLT database which we can use for remote queries.

Try connecting to the database through Azure Data Studio (Figure 4-14).

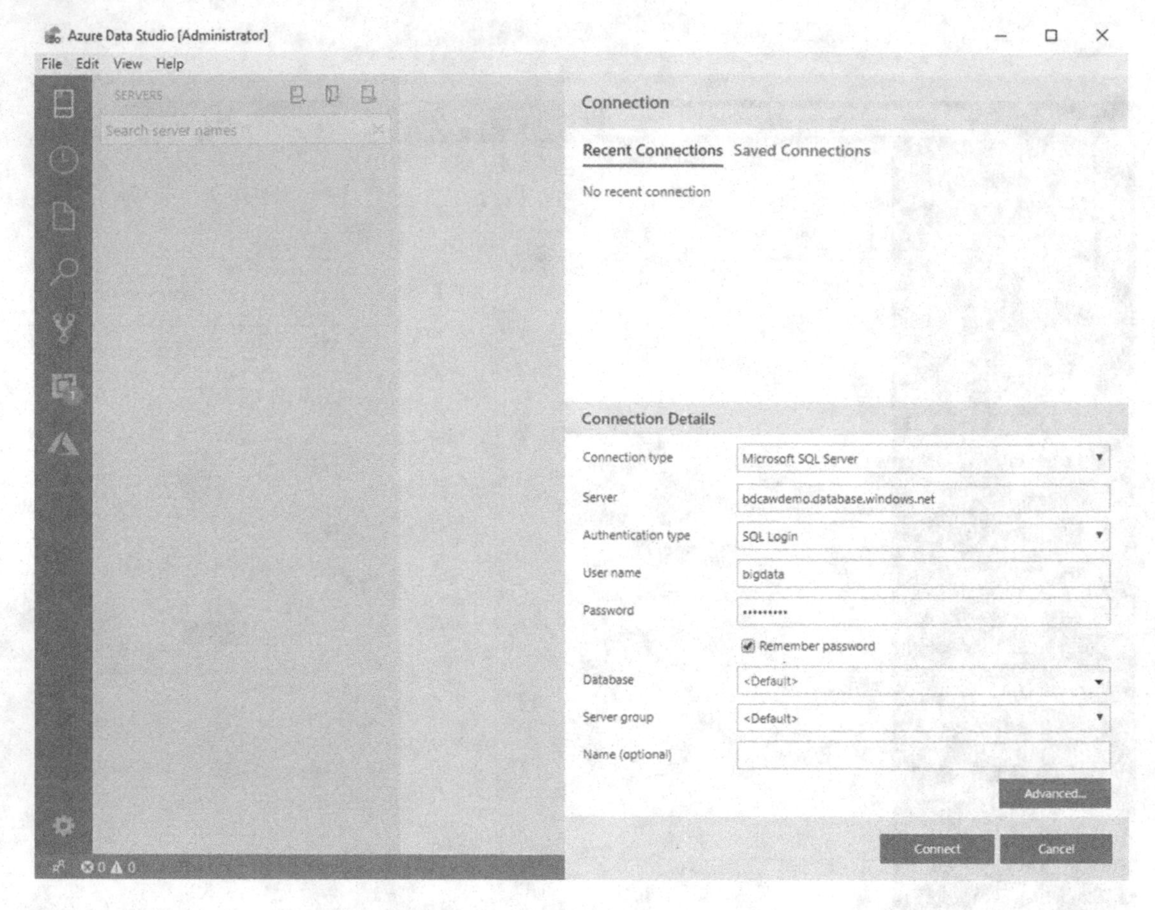


***Figure 4-14.** Connection dialog in ADS*

If you are not on an Azure VM or have unticked the box to allow connections from Azure, you will likely get the error shown in Figure 4-15.

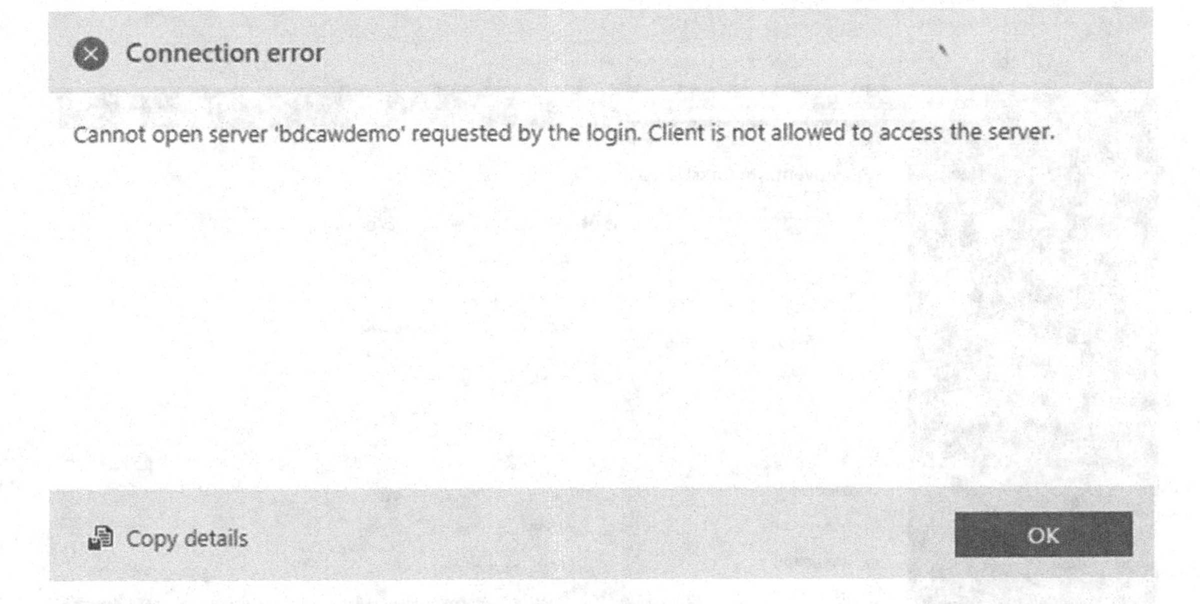


***Figure 4-15.*** *Azure Data Studio – Connection error*

If this happens, go back to the Azure Portal (Figure 4-16), navigate to your resource group containing the database server, and select the server (make sure to click the SQL Server, not the SQL Database).

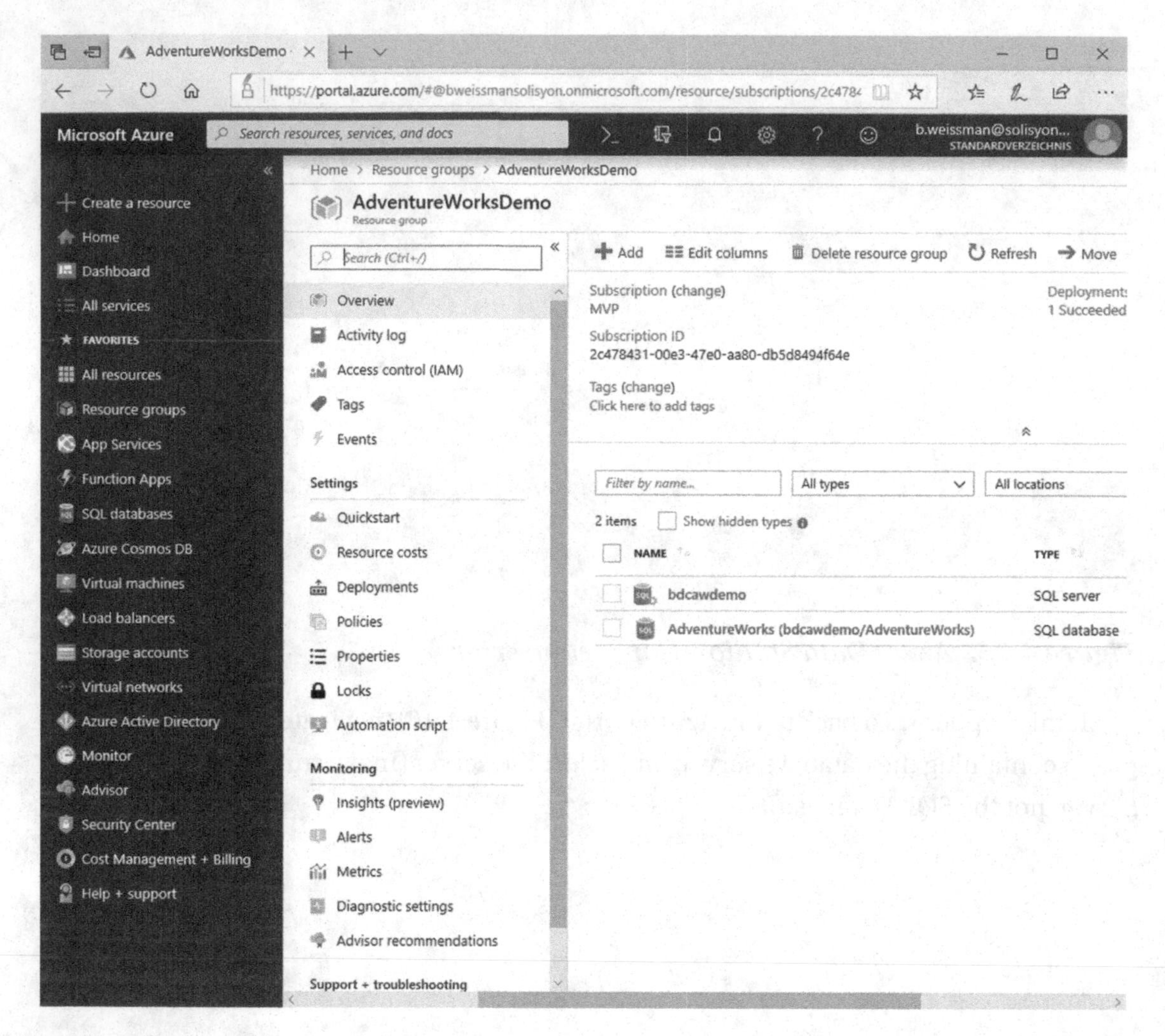


***Figure 4-16.** Azure SQL Database configuration*

On the left, scroll down to "Security" and pick "Firewalls and virtual networks" as shown in Figure 4-17.

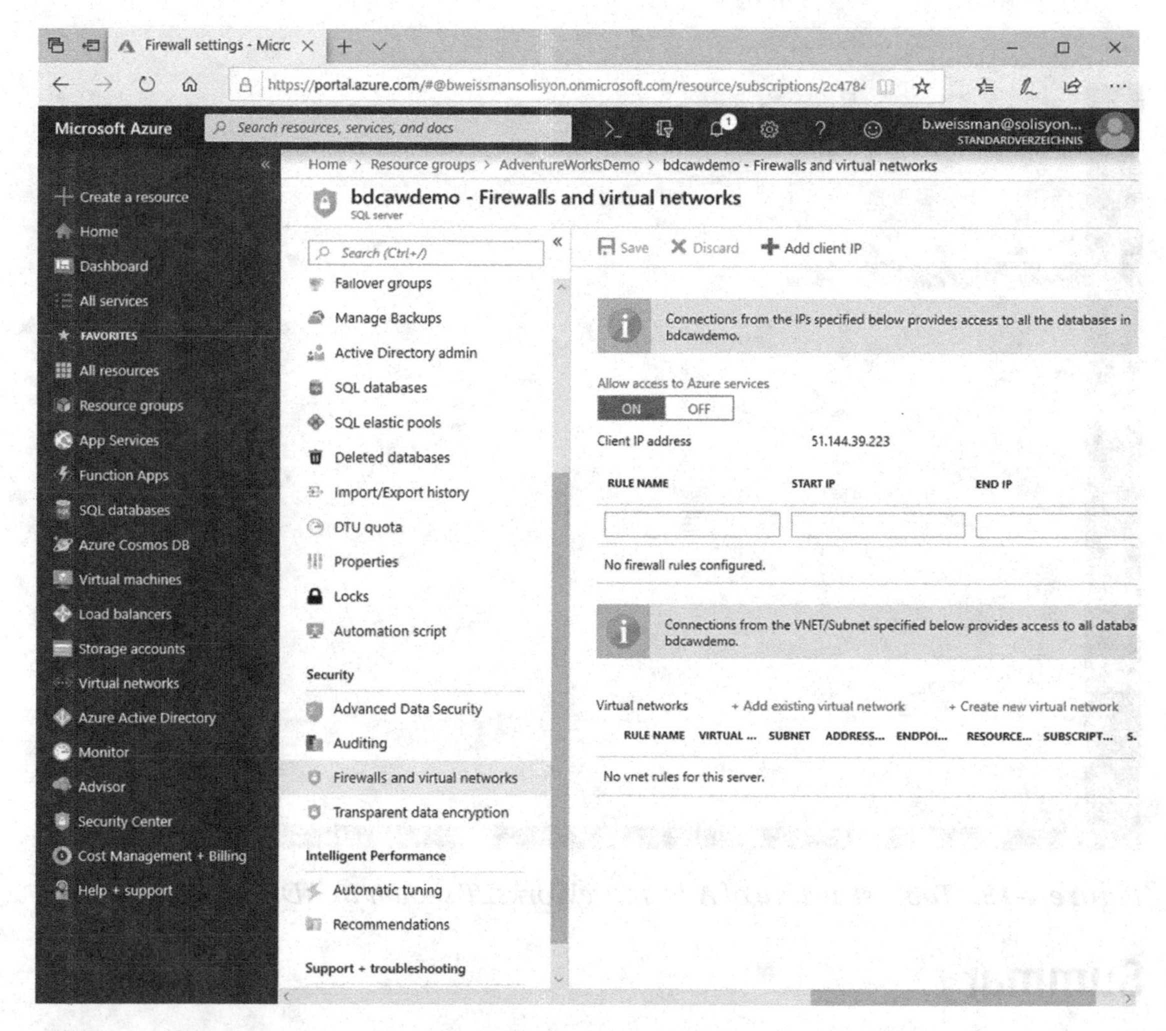


***Figure 4-17.*** *Azure Portal Firewall Settings*

Either click "Add client IP" or manually add your IP address.

Save your changes.

Now, try connecting to the database again in Azure Data Studio. You should be able to see the AdventureWorks Database including the tables shown in Figure 4-18.

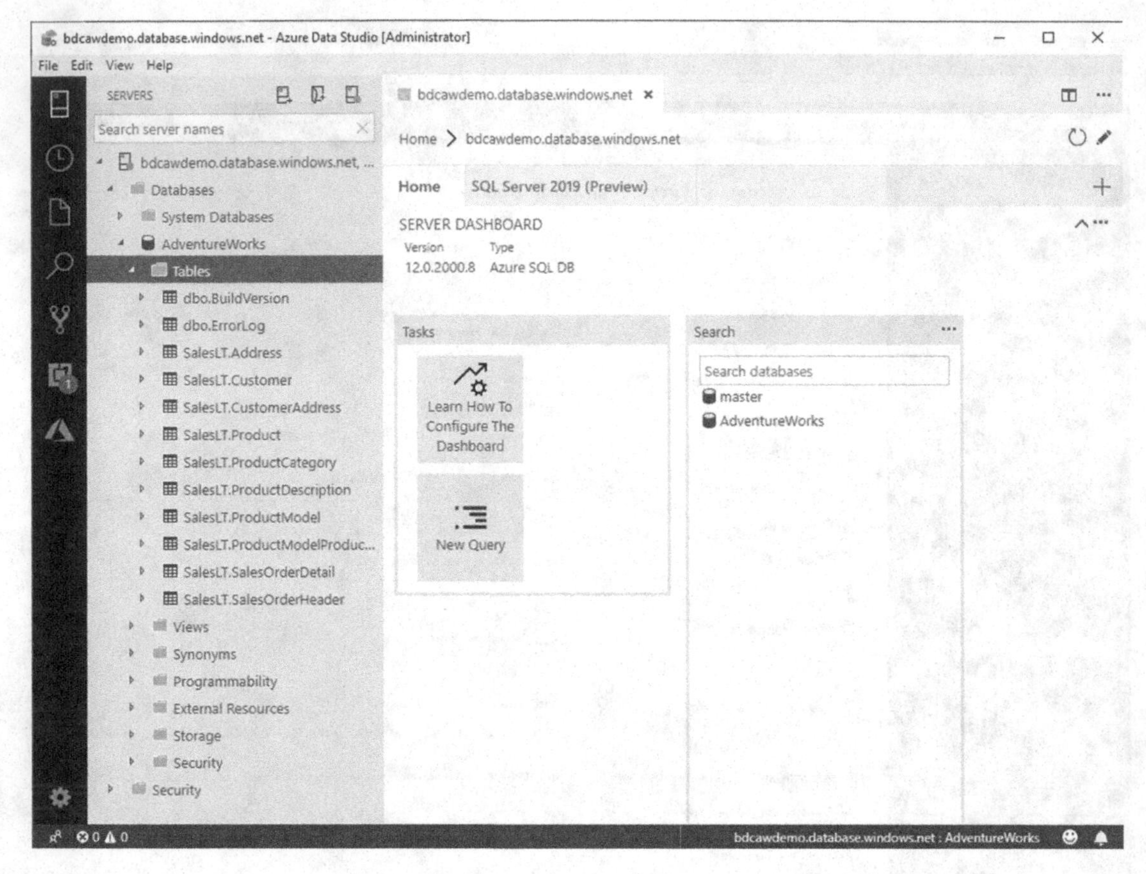


***Figure 4-18.** Table structure of AdventureWorksLT shown in ADS*

# Summary

In this chapter, we loaded some data into the previously deployed SQL Server Big Data Cluster. Now it's time to look on how to consume that data!

CHAPTER 5

# Querying Big Data Clusters Through T-SQL

Now that we have some data to play with, let's look at how we can process and query that data through the multiple options provided through Azure Data Studio.

## External Tables

Querying a Big Data Cluster using T-SQL happens through *external tables*, a concept that was introduced in SQL Server 2016 with the first appearance of PolyBase.

We will start to query our Big Data Cluster by adding some external tables to our new empty database BDC_Empty which originally resides in our AdventureWorksLT database in Azure.

To get started, connect to your SQL Server master instance (or any other SQL Server 2019 instance with PolyBase enabled) through Azure Data Studio as shown in Figure 5-1.




B. Weissman and E. van de Laar, *SQL Server Big Data Clusters*,
https://doi.org/10.1007/978-1-4842-5985-6_5

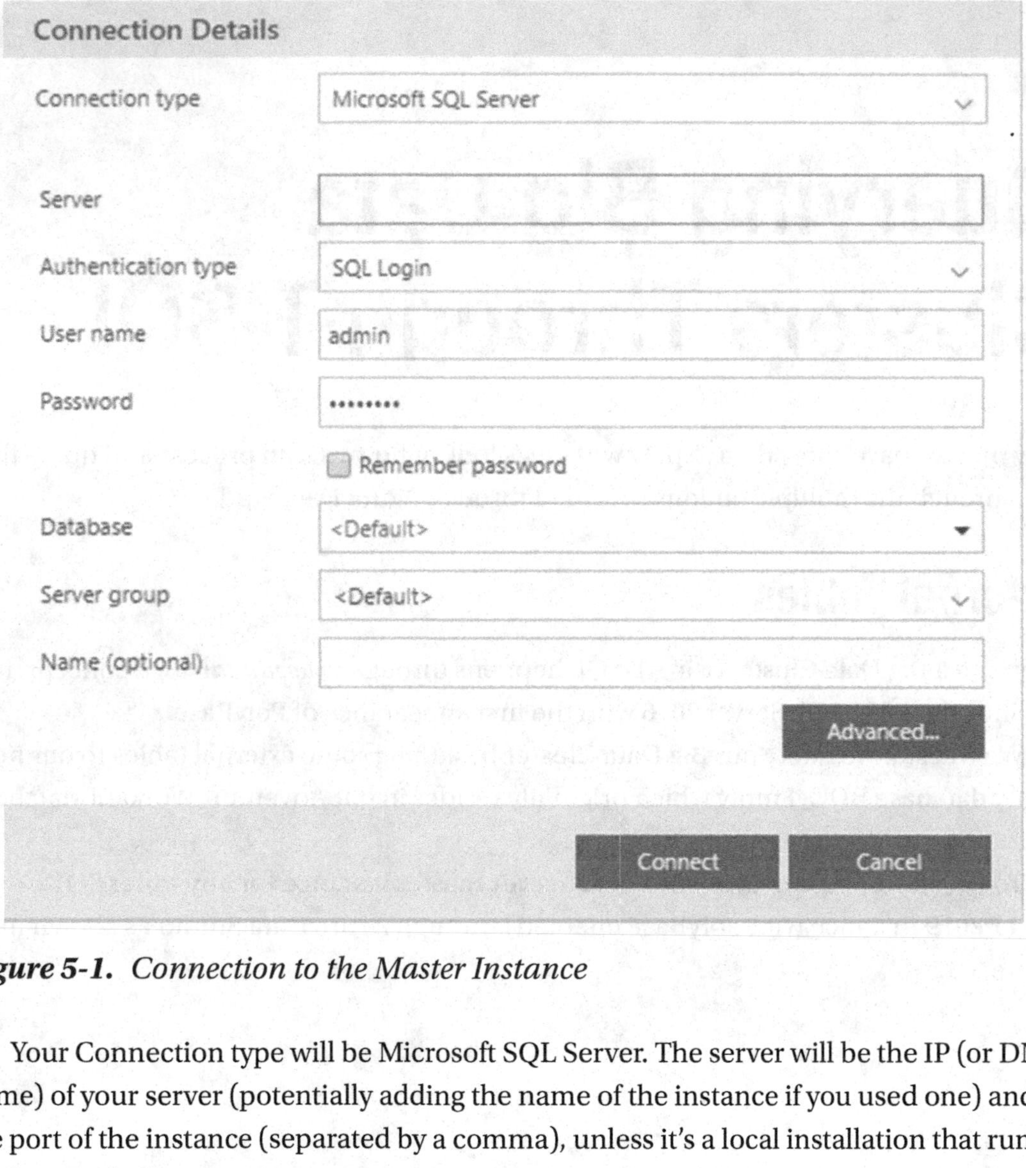

Connection Details

Connection type: Microsoft SQL Server
Server:
Authentication type: SQL Login
User name: admin
Password: ••••••••
Remember password
Database: <Default>
Server group: <Default>
Name (optional)
Advanced...
Connect
Cancel

***Figure 5-1.*** *Connection to the Master Instance*

Your Connection type will be Microsoft SQL Server. The server will be the IP (or DNS name) of your server (potentially adding the name of the instance if you used one) and the port of the instance (separated by a comma), unless it's a local installation that runs on the standard port 1433.

Expand your Connection, Databases, the BDC_Empty database, as well as the tables in it (Figure 5-2).

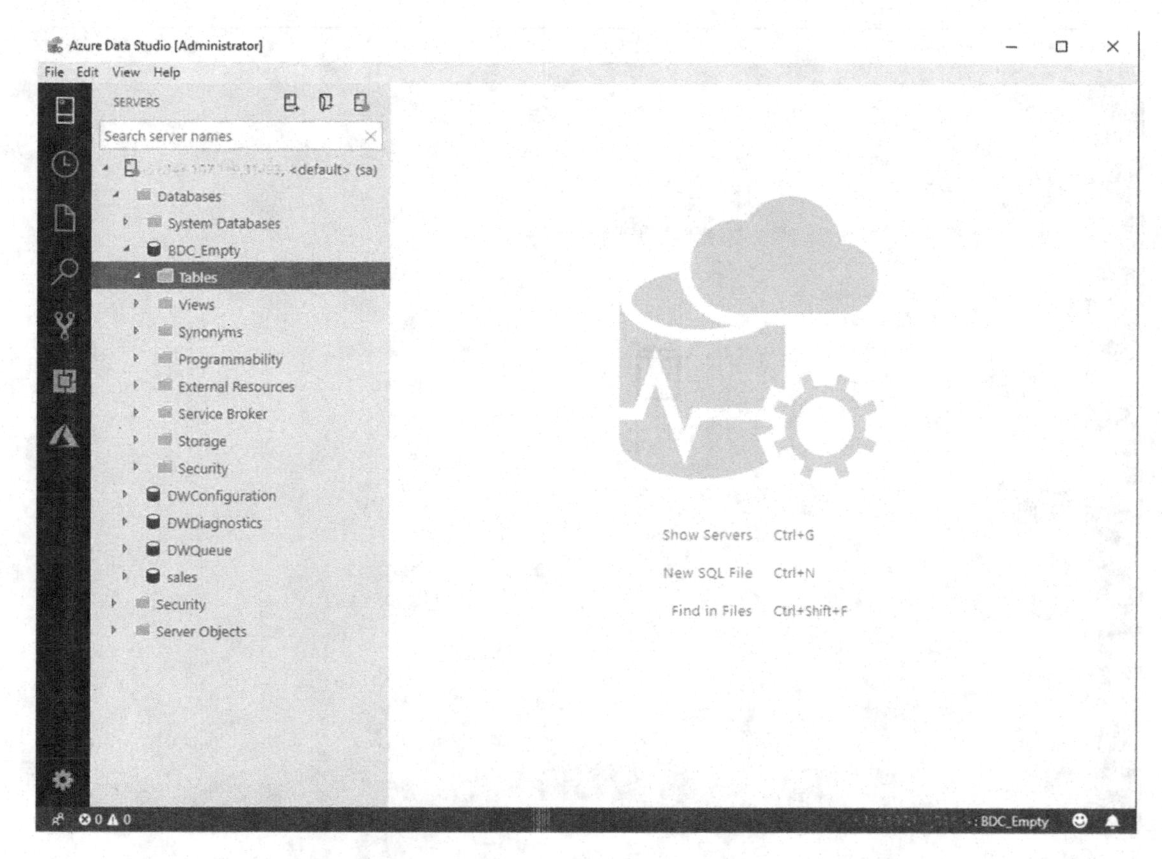


***Figure 5-2.** Empty database in ADS*

As expected, there are none so far. Let's change that!

If you right-mouse-click the database, you will see an option called "Create External Table." This will open up the corresponding wizard.

In the first step, it will ask you to confirm the database in which you want the external tables to be created as well as to choose a data source type. At this point, the wizard (Figure 5-3) supports SQL Server and Oracle; all other sources except CSV files (which have their own wizard) need to be scripted manually.

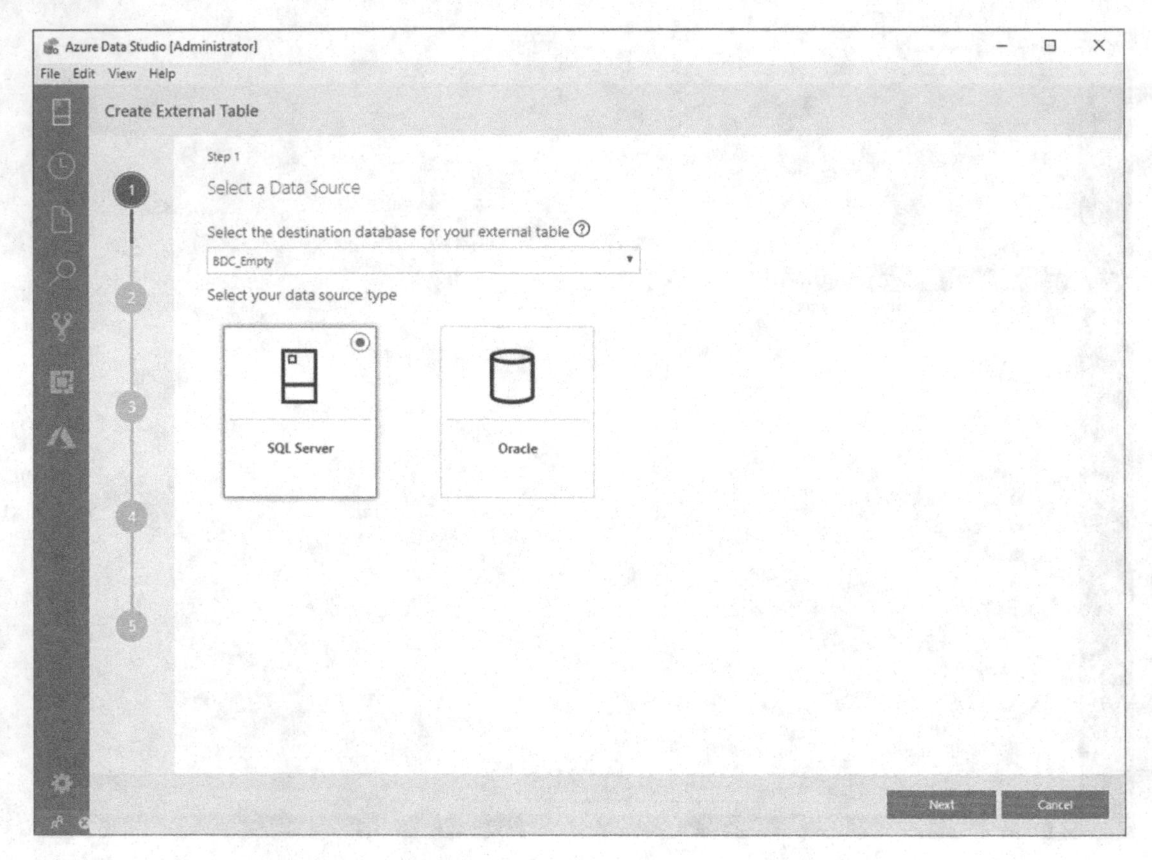


*Figure 5-3. External Table Wizard in ADS – Select a Data Source*

Select "SQL Server" and click Next.

In the next dialog (Figure 5-4), the wizard will ask you to set a master key password for this database. This is required, as we'll store credentials in the database which need to be encrypted. If you run the wizard on a database that already has a master key password, this step will be skipped.

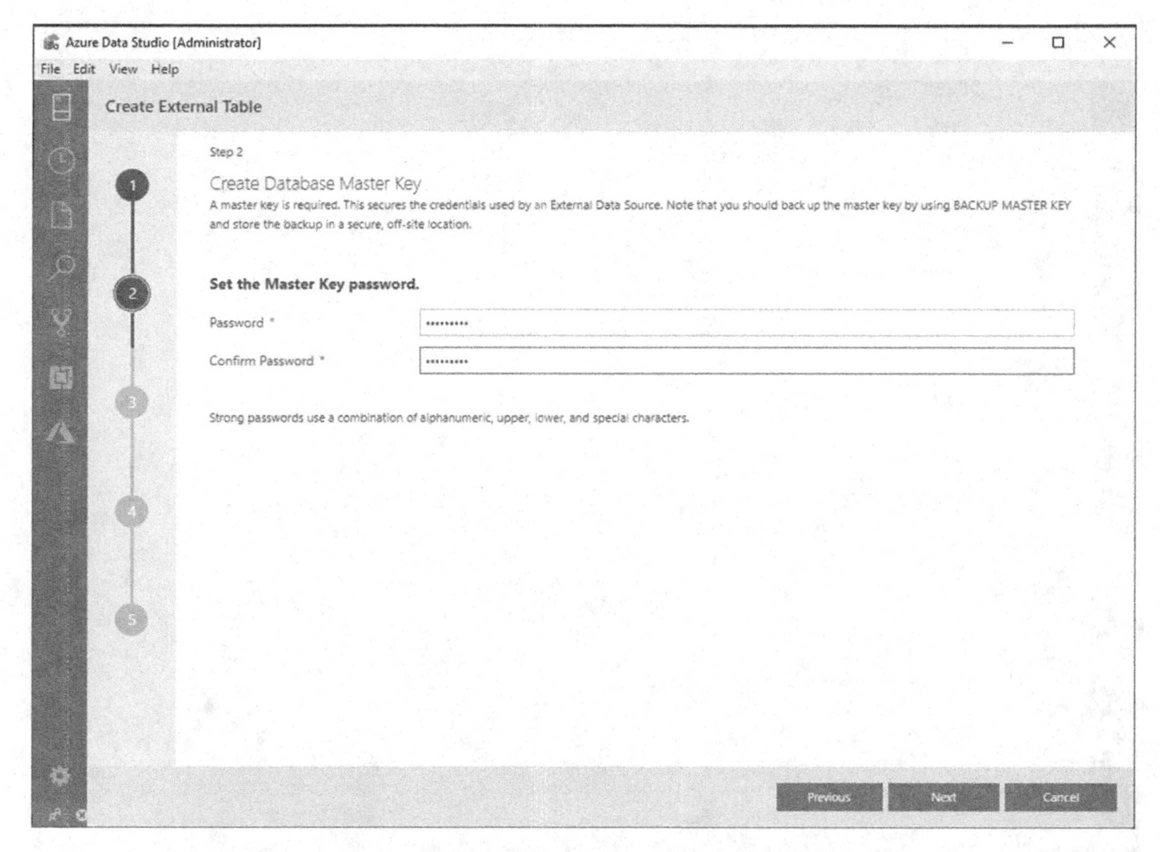


***Figure 5-4.** External Table Wizard in ADS – Create Database Master Key*

Enter and confirm a password and click Next.

The next screen (Figure 5-5) asks you for a name (an alias) for your connection, the connection's server name, as well as the database name. You can also use select an existing connection, if you've configured it previously.

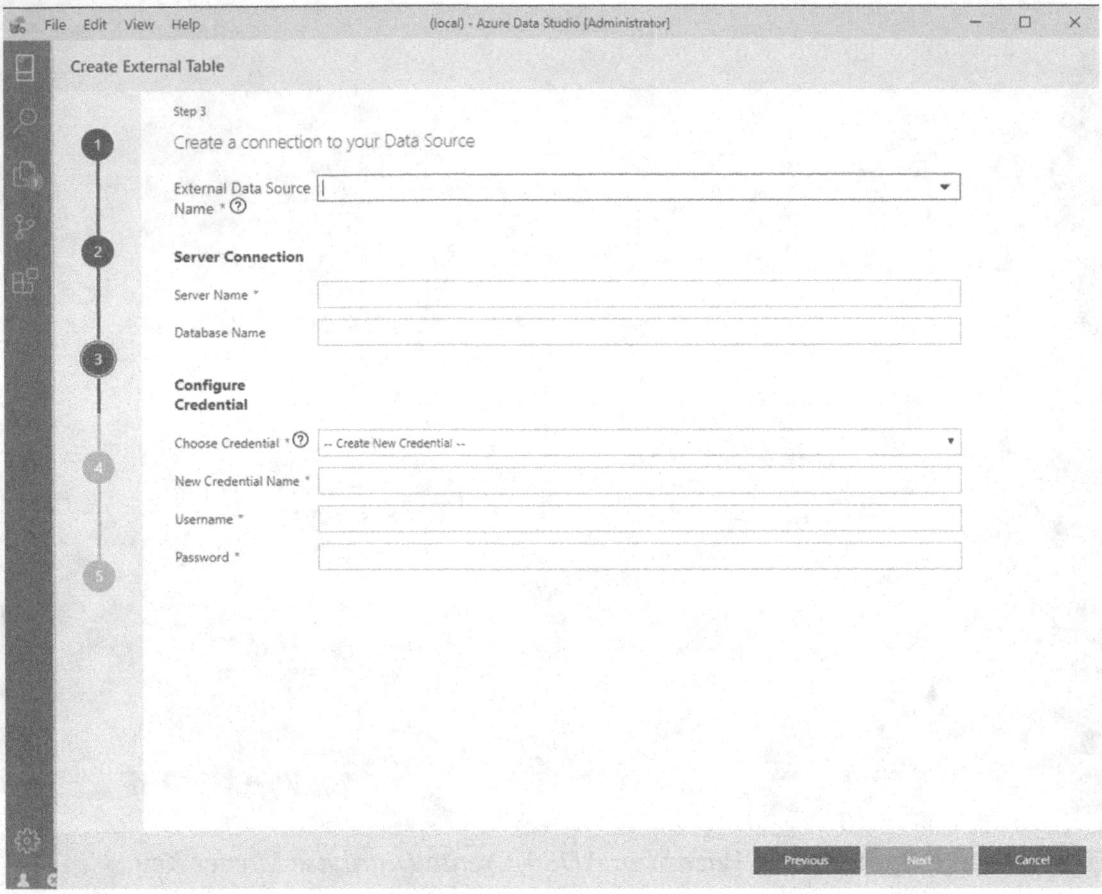

*Figure 5-5. External Table Wizard in ADS – connection and credentials*

In addition, you are provided with a list of credentials that already exist (if any) as well as the option to create a new credential.

Use "AW" for your data source name as well as the "New Credential Name," AdventureWorks as your Database Name, and provide the Server Name, Username, and Password that you've configured in the previous step. Click Next.

The wizard will now load all the tables and views in the source database which you can expand and browse as shown in Figure 5-6.

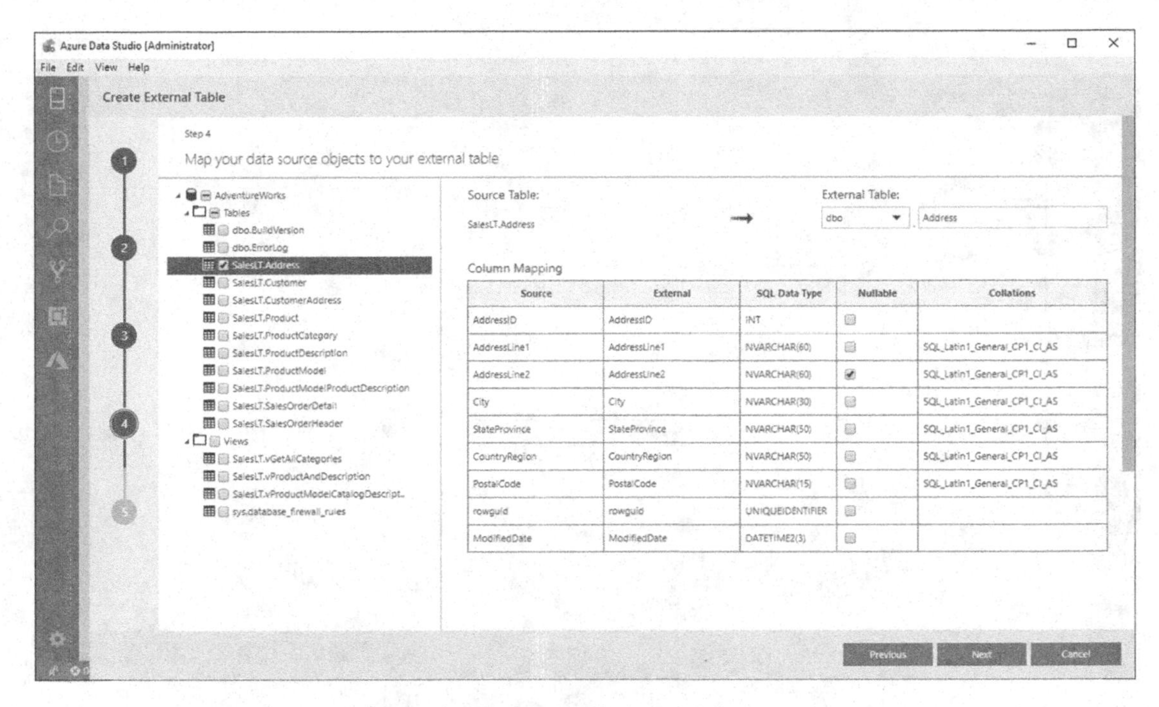


**Figure 5-6.** *External Table Wizard in ADS – object mapping*

You can either select the whole database, all tables or all views, or a variable number of single objects.

You won't be able to change any column definitions and you will always have to "create" all columns that exist within the source. As no data is actually moved but is only a reference to a foreign schema, this is not an issue.

The only two things that can be changed (in the upper-right section of the screen as shown in Figure 5-7) are the target schema and the target table name, as you may want to create all external tables in a separate schema or add a prefix to them.

**Figure 5-7.** *External Table Wizard in ADS – table mapping*

For now, pick the tables Address, Customer, and CustomerAddress and leave all other tables unticked as well as all settings unchanged. Click Next.

You have reached the last step, which is a summary as shown in Figure 5-8.

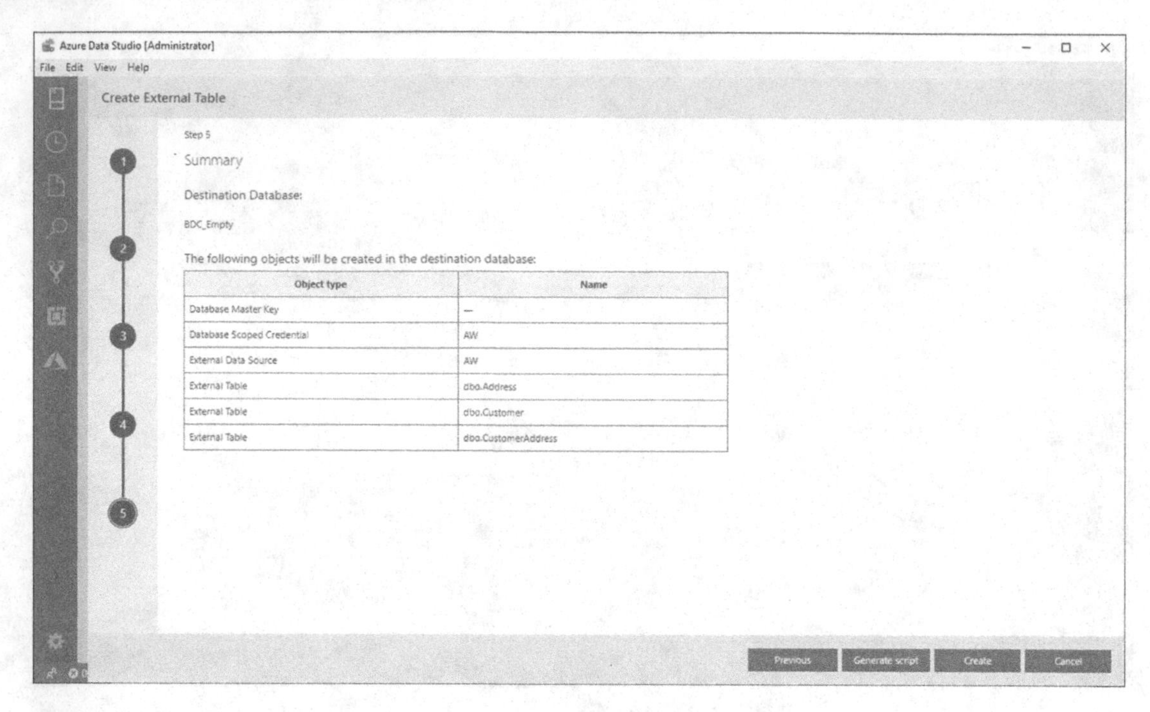


***Figure 5-8.*** *External Table Wizard in ADS - Summary*

You can now choose to either generate a script or simply create all Objects from the wizard.

Let us take a look at the script (Listing 5-1), so click "Generate Script" and then "Cancel" to close the wizard. You will see the script which will start a transaction and then create all the objects in the right order, starting with the key, followed by the credential and the data source and finally our three tables.

***Listing 5-1.*** T-SQL to generate external tables from Azure SQL DB

```
BEGIN TRY
    BEGIN TRANSACTION Tcfc2da095679401abd1ae9deb0e6eae
        USE [BDC_Empty];
        CREATE MASTER KEY ENCRYPTION BY PASSWORD = '<yourkey>';
        CREATE DATABASE SCOPED CREDENTIAL [AW]
            WITH IDENTITY = 'bigdata', SECRET = '<yourpassword>';
        CREATE EXTERNAL DATA SOURCE [AW]
            WITH (LOCATION = 'sqlserver:// <yourserver>.database.windows.
            net', CREDENTIAL = [AW]);
```

```
CREATE EXTERNAL TABLE [dbo].[Address]
(
    [AddressID] INT NOT NULL,
    [AddressLine1] NVARCHAR(60) COLLATE SQL_Latin1_General_CP1_CI_
     AS NOT NULL,
    [AddressLine2] NVARCHAR(60) COLLATE SQL_Latin1_General_CP1_CI_AS,
    [City] NVARCHAR(30) COLLATE SQL_Latin1_General_CP1_CI_AS NOT NULL,
    [StateProvince] NVARCHAR(50) COLLATE SQL_Latin1_General_CP1_CI_
     AS NOT NULL,
    [CountryRegion] NVARCHAR(50) COLLATE SQL_Latin1_General_CP1_CI_
     AS NOT NULL,
    [PostalCode] NVARCHAR(15) COLLATE SQL_Latin1_General_CP1_CI_AS
     NOT NULL,
    [rowguid] UNIQUEIDENTIFIER NOT NULL,
    [ModifiedDate] DATETIME2(3) NOT NULL
)
WITH (LOCATION = '[AdventureWorks].[SalesLT].[Address]',
DATA_SOURCE = [AW]);
CREATE EXTERNAL TABLE [dbo].[Customer]
(
    [CustomerID] INT NOT NULL,
    [NameStyle] BIT NOT NULL,
    [Title] NVARCHAR(8) COLLATE SQL_Latin1_General_CP1_CI_AS,
    [FirstName] NVARCHAR(50) COLLATE SQL_Latin1_General_CP1_CI_AS
     NOT NULL,
    [MiddleName] NVARCHAR(50) COLLATE SQL_Latin1_General_CP1_CI_AS,
    [LastName] NVARCHAR(50) COLLATE SQL_Latin1_General_CP1_CI_AS
     NOT NULL,
    [Suffix] NVARCHAR(10) COLLATE SQL_Latin1_General_CP1_CI_AS,
    [CompanyName] NVARCHAR(128) COLLATE SQL_Latin1_General_CP1_CI_AS,
    [SalesPerson] NVARCHAR(256) COLLATE SQL_Latin1_General_CP1_CI_AS,
    [EmailAddress] NVARCHAR(50) COLLATE SQL_Latin1_General_CP1_CI_AS,
    [Phone] NVARCHAR(25) COLLATE SQL_Latin1_General_CP1_CI_AS,
    [PasswordHash] VARCHAR(128) COLLATE SQL_Latin1_General_CP1_CI_
     AS NOT NULL,
```

```
        [PasswordSalt] VARCHAR(10) COLLATE SQL_Latin1_General_CP1_CI_AS
         NOT NULL,
        [rowguid] UNIQUEIDENTIFIER NOT NULL,
        [ModifiedDate] DATETIME2(3) NOT NULL
    )
    WITH (LOCATION = '[AdventureWorks].[SalesLT].[Customer]',
    DATA_SOURCE = [AW]);
    CREATE EXTERNAL TABLE [dbo].[CustomerAddress]
    (
        [CustomerID] INT NOT NULL,
        [AddressID] INT NOT NULL,
        [AddressType] NVARCHAR(50) COLLATE SQL_Latin1_General_CP1_CI_AS
         NOT NULL,
        [rowguid] UNIQUEIDENTIFIER NOT NULL,
        [ModifiedDate] DATETIME2(3) NOT NULL
    )
    WITH (LOCATION = '[AdventureWorks].[SalesLT].[CustomerAddress]',
    DATA_SOURCE = [AW]);
  COMMIT TRANSACTION Tcfc2da095679401abd1ae9deb0e6eae
END TRY
BEGIN CATCH
  IF @@TRANCOUNT > 0
    ROLLBACK TRANSACTION Tcfc2da095679401abd1ae9deb0e6eae
  DECLARE @ErrorMessage NVARCHAR(4000) = ERROR_MESSAGE();
  DECLARE @ErrorSeverity INT = ERROR_SEVERITY();
  DECLARE @ErrorState INT = ERROR_STATE();
  RAISERROR(@ErrorMessage, @ErrorSeverity, @ErrorState);
END CATCH;
```

Once you run that script (click Run on the upper-left part of the screen or simply hit F5), it will execute and create those objects in your database.

If you refresh your tables, the three new tables will show up and it will look like Figure 5-9. You can recognize that they are external tables easily by the hint behind the table names.

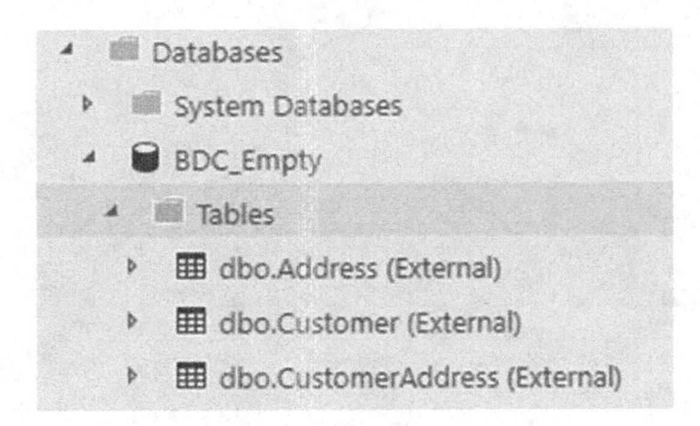


***Figure 5-9.*** *External tables shown after creation in ADS*

In SSMS, they can be recognized by sitting in their own folder instead as shown in Figure 5-10.

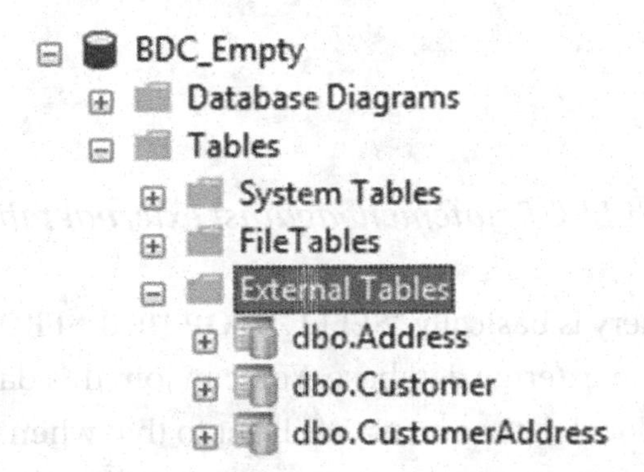


***Figure 5-10.*** *External tables shown after creation in SSMS*

From a client perspective, these tables behave like local tables. Make a right-mouse-click on the Address table in Azure Data Studio (Figure 5-11) and click "Select Top 1000."

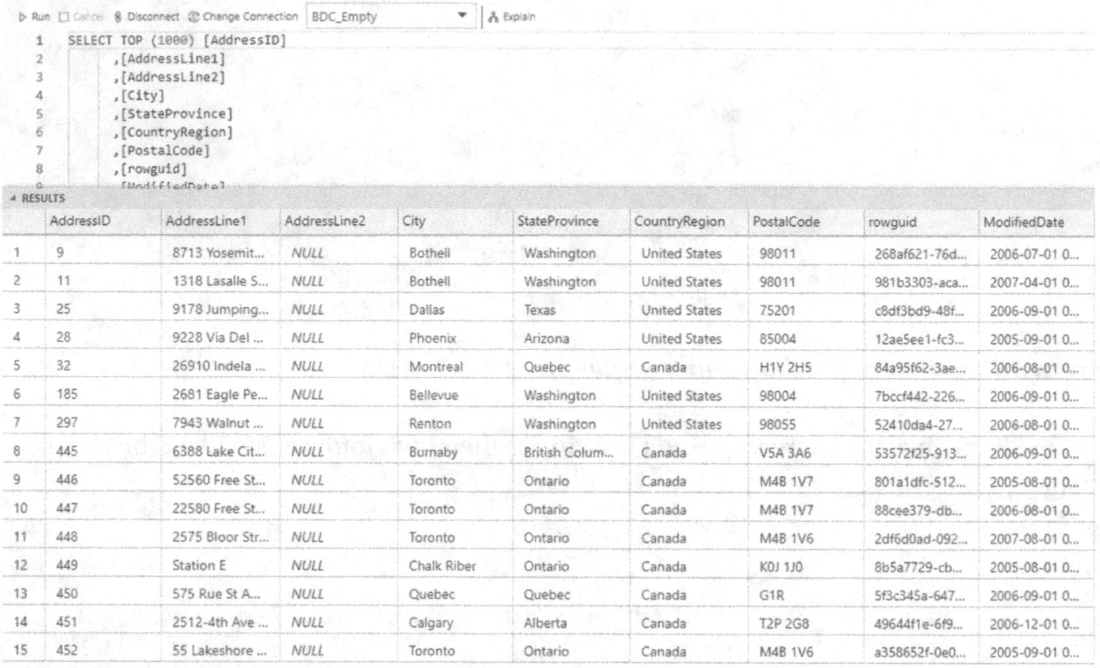

Run Cancel Disconnect Change Connection BDC_Empty Explain

```
1  SELECT TOP (1000) [AddressID]
2        ,[AddressLine1]
3        ,[AddressLine2]
4        ,[City]
5        ,[StateProvince]
6        ,[CountryRegion]
7        ,[PostalCode]
8        ,[rowguid]
```

RESULTS

| | AddressID | AddressLine1 | AddressLine2 | City | StateProvince | CountryRegion | PostalCode | rowguid | ModifiedDate |
|---|---|---|---|---|---|---|---|---|---|
| 1 | 9 | 8713 Yosemit... | NULL | Bothell | Washington | United States | 98011 | 268af621-76d... | 2006-07-01 0... |
| 2 | 11 | 1318 Lasalle S... | NULL | Bothell | Washington | United States | 98011 | 981b3303-aca... | 2007-04-01 0... |
| 3 | 25 | 9178 Jumping... | NULL | Dallas | Texas | United States | 75201 | c8df3bd9-48f... | 2006-09-01 0... |
| 4 | 28 | 9228 Via Del ... | NULL | Phoenix | Arizona | United States | 85004 | 12ae5ee1-fc3... | 2005-09-01 0... |
| 5 | 32 | 26910 Indela ... | NULL | Montreal | Quebec | Canada | H1Y 2H5 | 84a95f62-3ae... | 2006-08-01 0... |
| 6 | 185 | 2681 Eagle Pe... | NULL | Bellevue | Washington | United States | 98004 | 7bccf442-226... | 2006-09-01 0... |
| 7 | 297 | 7943 Walnut ... | NULL | Renton | Washington | United States | 98055 | 52410da4-27... | 2006-08-01 0... |
| 8 | 445 | 6388 Lake Cit... | NULL | Burnaby | British Colum... | Canada | V5A 3A6 | 53572f25-913... | 2006-09-01 0... |
| 9 | 446 | 52560 Free St... | NULL | Toronto | Ontario | Canada | M4B 1V7 | 801a1dfc-512... | 2005-08-01 0... |
| 10 | 447 | 22580 Free St... | NULL | Toronto | Ontario | Canada | M4B 1V7 | 88cee379-db... | 2006-08-01 0... |
| 11 | 448 | 2575 Bloor Str... | NULL | Toronto | Ontario | Canada | M4B 1V6 | 2df6d0ad-092... | 2007-08-01 0... |
| 12 | 449 | Station E | NULL | Chalk Riber | Ontario | Canada | K0J 1J0 | 8b5a7729-cb... | 2005-08-01 0... |
| 13 | 450 | 575 Rue St A... | NULL | Quebec | Quebec | Canada | G1R | 5f3c345a-647... | 2006-09-01 0... |
| 14 | 451 | 2512-4th Ave ... | NULL | Calgary | Alberta | Canada | T2P 2G8 | 49644f1e-6f9... | 2006-12-01 0... |
| 15 | 452 | 55 Lakeshore ... | NULL | Toronto | Ontario | Canada | M4B 1V6 | a358652f-0e0... | 2005-09-01 0... |

***Figure 5-11.*** *Output of SELECT statement against external table in ADS*

You can see that the query is basically "SELECT TOP 1000 * FROM dbo.Address", despite that data sitting in an external database. You can join this data against local tables or any other kind of local data sources. We'll get to that when we look at external tables from CSV files.

Let us start by running a query (Listing 5-2) against all three external tables to get all companies whose main office is in Idaho.

***Listing 5-2.*** SELECT statement joining two external tables

```
SELECT CompanyName
  FROM [Address] ADDR
  INNER JOIN CustomerAddress CADDR ON ADDR.AddressID = CADDR.AddressID
  INNER JOIN Customer CUST ON CUST.CustomerID = CADDR.CustomerID
  WHERE
    AddressType = 'Main Office'
    AND StateProvince = 'Idaho'
```

Again, this looks like a regular query on some local tables as shown in Figure 5-12.

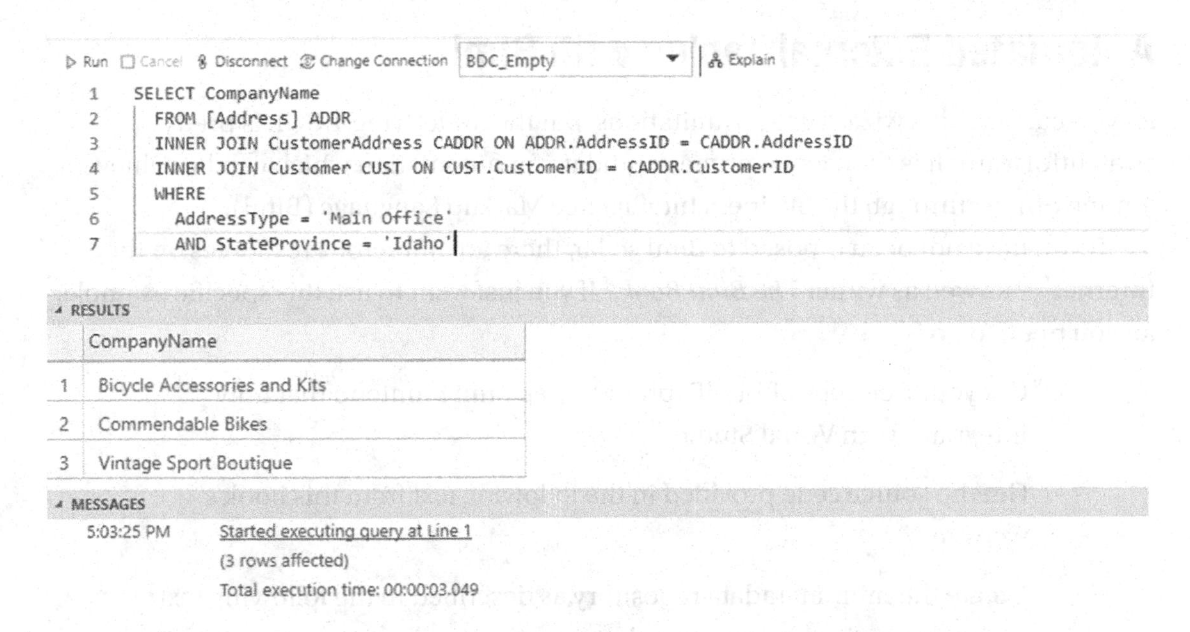


***Figure 5-12.** Output of joined SELECT statement in ADS*

Only, when you click the Explain button on the upper right, you will see the execution plan (Figure 5-13), which reveals the fact that the query is running remote (Remote Query and External Select operators).

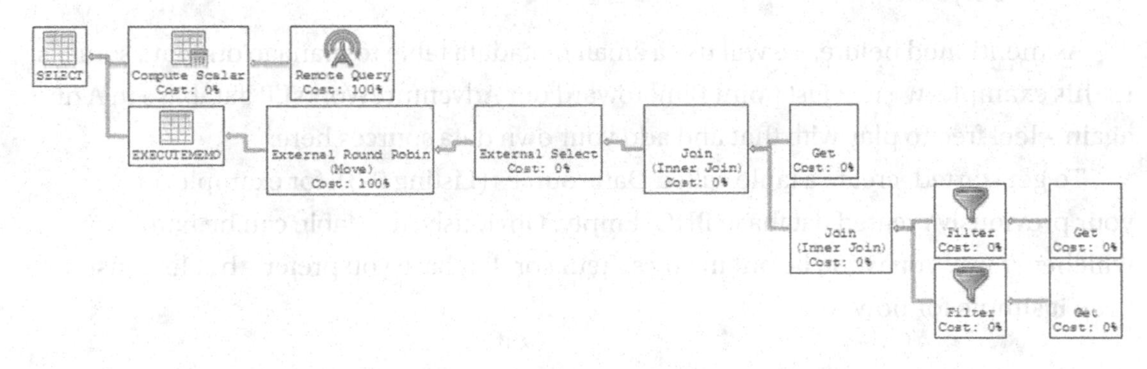


***Figure 5-13.** Execution plan against two external tables in ADS*

## Automated External Tables with Biml

As you can see, this wizard has its limitations, but the underlying T-SQL is pretty straightforward. It is therefore a prime candidate for automation which can be achieved, among others, through the Business Intelligence Markup Language (Biml).

If you have not been exposed to Biml so far, there are numerous resources on the Internet[1,2,3] as well as within *The Biml Book*.[4] If you just want to use this specific example, all you need to do is

- Get your free copy of BimlExpress,[5] a free Biml Frontend that fully integrates with Visual Studio.
- Get the source code provided in the following text from this book's website.
- Create the mini-metadata repository as described in the following text and populate it with your metadata.
- Adjust the connection strings in the solution.
- Run the solution (just right-mouse-click "11_Polybase_C.biml" in the solution and select "Generate SSIS Package"). Despite the confusing title, this will write all the required .SQL files to C:\Temp\Polybase.

As mentioned before, we will use a small metadata table to manage our data sources. In this example, we will just point Biml toward our AdventureWorksLT database in Azure again – feel free to play with that and add your own data sources here.

To get started, create a table called Datasources (Listing 5-3), for example, in your previously created database BDC_Empty. Obviously, this table can be named in whichever way you want, be put in any schema or database you prefer – but let's just keep it simple for now.

[1] www.cathrinewilhelmsen.net/biml/
[2] http://bimlscript.com/
[3] http://biml.blog/
[4] www.apress.com/de/book/9781484231340
[5] www.varigence.com/BimlExpress

***Listing 5-3.*** T-SQL to create datasources metatable

```
CREATE TABLE [dbo].[Datasources](
    [DataSource] [nvarchar](50) NULL,
    [Server] [nvarchar](50) NULL,
    [UserID] [nvarchar](50) NULL,
    [Password] [nvarchar](50) NULL,
    [SRC_DB] [nvarchar](50) NULL,
    [SRC_Schema] [nvarchar](50) NULL,
    [DEST_Schema] [nvarchar](50) NULL
) ON [PRIMARY]
```

Also, just add one record to this table pointing to your AWLT database (modify your DNS name, username, and password as needed) using the code in (Listing 5-4).

***Listing 5-4.*** Populate your datasources metatable

```
INSERT INTO Datasources VALUES
('AW','<yourinstance>.database.windows.net','<yourUser>','<yourPassword>','
AdventureWorks','SalesLT','dbo')
```

This solution expects the master key to be set up already - this being a one-time task, there is simply no need or justification to automate that step. If you skipped the manual table creation in the previous exercise, you may want to fulfill this step manually before continuing.

Our solution contains two Biml files (actually four - two files in C# and VB.NET, respectively):

- 11_Polybase_C.biml

  This is the control file which will hold the connection strings to our metadata as well as the target database. It will loop through the metadata and call the other Biml file (using a function called CallBimlScript) for every single entry, writing a .SQL file to C:\temp\polybase (so if you have ten entries in your Datasources table, you will end up with ten files).

- 12_PolybaseWriter_C.biml

  This file will generate the contents of each source schema.

If you take a look at the first file, you will notice that it starts with the declaration of two (in this case identical) connection strings. One points to your database that holds your metadata; the other one points to the PolyBase database in which you want the external tables to be created or updated (Listing 5-5).

***Listing 5-5.*** Biml code to loop over the datasources metatable

```
<#@ import namespace="System.Data"#>
<# string MetaString = "Data Source=.;Initial Catalog=BDC_Empty;Provider=
SQLNCLI11.1;Integrated Security=SSPI;";
string TargetString = "Data Source=.;Initial Catalog=BDC_Empty;Provider=
SQLNCLI11.1;Integrated Security=SSPI;";
DataTable ExternalConnections = ExternalDataAccess.
GetDataTable(MetaString,"SELECT * FROM Datasources");
foreach (DataRow conn in ExternalConnections.Rows) {
System.IO.File.WriteAllText(@"C:\Temp\Polybase\" + conn ["datasource"] +
"_" + conn["SRC_Schema"] + ".sql",  CallBimlScript("12_PolybaseWriter_C.
biml",conn,TargetString));
} #>
<Biml xmlns="http://schemas.varigence.com/biml.xsd"/>
```

The actual magic happens in the second file. It takes the DataRow containing the metadata as well as the connection string from the target database as its parameters. It will then generate T-SQL to

- CREATE or ALTER the credentials for the connection
- CREATE or ALTER the external data source
- DROP every existing external table in the target database
- CREATE a corresponding external table for every single table in the source scheme

For the first three steps, it uses simple SQL Selects or semistatic T-SQL. For the fourth part, it makes use of Biml's ability to read and interpret a database's schema and store it in the Biml object model (Listing 5-6).

***Listing 5-6.*** Biml "12_PolybaseWriter_C.biml" called by previous Biml script

```
<#@ import namespace="Varigence.Biml.CoreLowerer.TSqlEmitter" #>
<#@ import namespace="System.Data" #>
<#@ property name="conn" type="DataRow" #>
<#@ property name="TargetString" type="String" #>
-- Syncing schema <#= conn["SRC_Schema"] #> in <#= conn["SRC_DB"]#> to
   <#= conn["DEST_Schema"]#>
-- This script assumes that a master key has been set
-- CREATE/ALTER CREDENTIAL
IF NOT EXISTS(select * from sys.database_credentials WHERE NAME =
'<#= conn["DataSource"]#>')
BEGIN
CREATE DATABASE SCOPED CREDENTIAL [<#= conn["DataSource"]#>]
       WITH IDENTITY = '<#= conn["UserID"]#>', SECRET = '<#= conn["Password"]#>';
END
ELSE
BEGIN
ALTER DATABASE SCOPED CREDENTIAL [<#= conn["DataSource"]#>]
       WITH IDENTITY = '<#= conn["UserID"]#>', SECRET =
'<#= conn["Password"]#>';
END
GO
-- CREATE DATASOURCE
IF NOT EXISTS(SELECT * FROM sys.external_data_sources WHERE NAME =
'<#= conn["DataSource"]#>')
BEGIN
CREATE EXTERNAL DATA SOURCE [<#= conn["DataSource"]#>]
            WITH (LOCATION = 'sqlserver://<#= conn["Server"]#>',
            CREDENTIAL = [<#= conn["DataSource"]#>]);
END
ELSE
BEGIN
ALTER EXTERNAL DATA SOURCE [<#= conn["DataSource"]#>] SET LOCATION =
N'sqlserver://<#= conn["Server"]#>', CREDENTIAL = [<#= conn["DataSource"]#>]
END
```

```
GO
-- DROP EXISTING TABLES
<# string DropSQL = "SELECT schem.name SchemaName,tbl.Name TableName,object_id
FROM sys.external_tables tbl  INNER JOIN sys.schemas schem on tbl.schema_id =
schem.schema_id INNER JOIN sys.external_data_sources ds on tbl.data_source_id =
ds.data_source_id WHERE ds.name = '" + conn["DataSource"] + "'";DataTable
ExistingTables = ExternalDataAccess.GetDataTable(TargetString,DropSQL);
foreach (DataRow tbl in ExistingTables.Rows) { #>
IF EXISTS(select * from sys.external_tables WHERE object_id = <#=
tbl["object_id"]#>)
BEGIN
DROP EXTERNAL TABLE [<#= tbl["SchemaName"]#>].[<#= tbl["TableName"]#>]
END
GO
<# } #>
-- CREATE TABLES
<# string Src_ConnStr= "Data Source=" + conn["Server"] + ";Initial Catalog="
+ conn["SRC_DB"] + ";Provider=SQLNCLI11.1;user id=" + conn["UserID"] +
";Password=" + conn["password"] + ";";
string SRC_Schema = conn["SRC_Schema"] + "";
var srcMeta = SchemaManager.CreateConnectionNode("Source", Src_ConnStr).
ImportDB(SRC_Schema,null,ImportOptions.None);
foreach (AstTableNode tbl in srcMeta.TableNodes) {
    foreach (AstTableColumnNode col in tbl.Columns.Where(c => c.DataType ==
    DbType.Xml)) {
        col.DataType = DbType.AnsiString;
        col.Length = 8000;
    }
    foreach (AstTableColumnNode col in tbl.Columns.Where(c => (c.DataType
    == DbType.String) && (c.Length == -1))) {
        col.Length = 4000;
    }
    foreach (AstTableColumnNode col in tbl.Columns.Where(c => (c.DataType ==
    DbType.AnsiString || c.DataType ==  DbType.Binary) & c.Length == -1)) {
        col.Length = 8000;
    } #>
```

```
IF NOT EXISTS(SELECT * FROM sys.external_tables WHERE NAME = '<#=tbl.Name#>')
BEGIN
CREATE EXTERNAL TABLE [<#= conn["DEST_Schema"] #>].[<#=tbl.Name#>] (
<#=string.Join(",\n",tbl.Columns.Select(i => i.Name + " " +
TSqlTypeTranslator.Translate(i.DataType, i.Length, i.Precision, i.Scale,
i.CustomType) + (i.IsNullable ? " NULL" : " NOT NULL")))#>
)
WITH (LOCATION = '[<#= conn["SRC_DB"] #>].<#= tbl.SchemaQualifiedName#>',
DATA_SOURCE = [<#= conn["DataSource"] #>]);
END
GO
<# } #>
```

As the focus of this book is Big Data Clusters, not Biml, we won't go into any more details of this little helper. The main idea was to show you one of multiple ways on how you can automate your way through external tables.

By the way, it would be super easy to adjust this code to work for other relational sources such as Teradata or Oracle and automate external tables on these as well!

## External Tables from CSV Files in HDFS

As you've learned already, besides other relational databases, you can also query flat files using T-SQL and PolyBase. As flat files do not have a "one-size-fits-all" format due to different delimiters and so on, we need to define at least one format definition. This definition resides in your database, so the same definition can be shared by multiple files, but you need to re-create the definition for every database you want the format to be available in.

Let's start with a simple example (Listing 5-7) in the sales database that was included in the Microsoft samples.

***Listing 5-7.*** T-SQL code to create external file format

```
CREATE EXTERNAL FILE FORMAT csv_file
WITH (
    FORMAT_TYPE = DELIMITEDTEXT,
    FORMAT_OPTIONS(
        FIELD_TERMINATOR = ',',
```

```
        STRING_DELIMITER = '"',
        FIRST_ROW = 2,
        USE_TYPE_DEFAULT = TRUE)
);
```

This T-SQL code will create a format called csv_file; the file will be a delimited text file with double quotes as your text qualifier and a comma as your delimiter. The first row will be skipped. The parameter USE_TYPE_DEFAULT will determine how to handle missing fields. If it's false, a missing field in the file will be NULL; otherwise, it will be 0 for numeric, an empty string for character-based columns, and 01/01/1900 for any date columns.

In this case, we will use the *StoragePool* which is the HDFS storage that is built into Big Data Clusters. To be able to access the StoragePool, you will need to create an external data source that points to it as shown in Listing 5-8.

***Listing 5-8.*** T-SQL code to create pointer to the storage pool

```
IF NOT EXISTS(SELECT * FROM sys.external_data_sources WHERE name =
'SqlStoragePool')
    CREATE EXTERNAL DATA SOURCE SqlStoragePool
    WITH (LOCATION = 'sqlhdfs://controller-svc/default');
```

With the format and data source in place, we can now create an external table that references this format as well as a file location (Listing 5-9).

***Listing 5-9.*** T-SQL code to create an external table based on a CSV file

```
CREATE EXTERNAL TABLE [web_clickstreams_hdfs_csv]
("wcs_click_date_sk" BIGINT , "wcs_click_time_sk" BIGINT , "wcs_sales_sk"
BIGINT , "wcs_item_sk" BIGINT , "wcs_web_page_sk" BIGINT , "wcs_user_sk"
BIGINT)
WITH
(
    DATA_SOURCE = SqlStoragePool,
        LOCATION = '/clickstream_data',
    FILE_FORMAT = csv_file
);
```

Just as with the SQL Server–based external tables, the first step is to define the columns including their names and datatypes. In addition, we need to provide a DATA_SOURCE which is our SqlStoragePool, so basically the HDFS of our Big Data Cluster (where we also uploaded the flight delay samples earlier); a LOCATION within that source (in this case, the clickstream_data subfolder); and a FILE_FORMAT which is our csv_file format we've created in the previous step.

Again, no data is transferred at this point. All we did was to create references to data residing somewhere else – in this case within the storage pool.

We can now live-query this file by a simple query like this (Listing 5-10).

***Listing 5-10.*** SELECT statement against csv-based external table

```
SELECT * FROM [dbo].[web_clickstreams_hdfs_csv]
```

But we can also join the data from the CSV with data that sits in a regular table within this database (Listing 5-11).

***Listing 5-11.*** SELECT statement joining a regular table with a csv-based external table

```
SELECT
    wcs_user_sk,
    SUM( CASE WHEN i_category = 'Books' THEN 1 ELSE 0 END) AS book_category_clicks,
    SUM( CASE WHEN i_category_id = 1 THEN 1 ELSE 0 END) AS [Home & Kitchen],
    SUM( CASE WHEN i_category_id = 2 THEN 1 ELSE 0 END) AS [Music],
    SUM( CASE WHEN i_category_id = 3 THEN 1 ELSE 0 END) AS [Books],
    SUM( CASE WHEN i_category_id = 4 THEN 1 ELSE 0 END) AS [Clothing &
    Accessories],
    SUM( CASE WHEN i_category_id = 5 THEN 1 ELSE 0 END) AS [Electronics],
    SUM( CASE WHEN i_category_id = 6 THEN 1 ELSE 0 END) AS [Tools & Home
    Improvement],
    SUM( CASE WHEN i_category_id = 7 THEN 1 ELSE 0 END) AS [Toys & Games],
    SUM( CASE WHEN i_category_id = 8 THEN 1 ELSE 0 END) AS [Movies & TV],
    SUM( CASE WHEN i_category_id = 9 THEN 1 ELSE 0 END) AS [Sports & Outdoors]
  FROM [dbo].[web_clickstreams_hdfs_csv]
  INNER JOIN item it ON (wcs_item_sk = i_item_sk
                        AND wcs_user_sk IS NOT NULL)
GROUP BY  wcs_user_sk;
```

Let's take a look at the execution plan of this query (Figure 5-14).

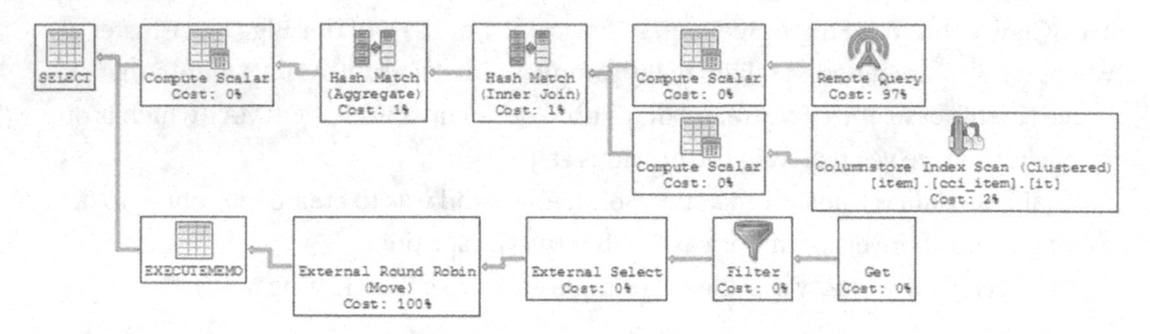


***Figure 5-14.*** *Execution plan of previous SELECT statement in ADS*

As you can see, this is – as expected – a combination of well-known operations like a clustered columnstore index scan as well as new features like the external select which are eventually merged together.

Of course, we can also join data from multiple CSVs. Therefore, we create an external table for each of our flight delay CSV files first. To help with that, there is another wizard (Figure 5-15).

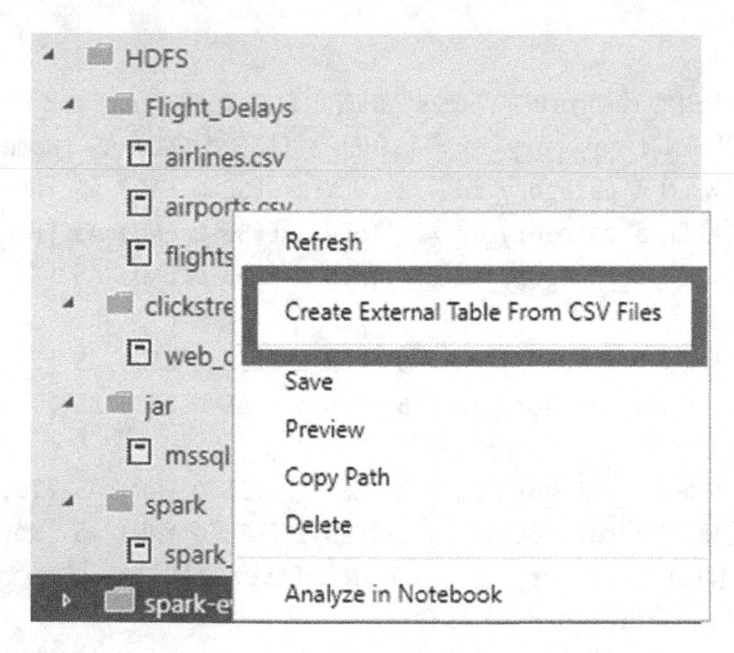


***Figure 5-15.*** *Create external table from CSV menu in ADS*

Right-click the file "airlines.csv" in Azure Data Studio and select "Create External Table From CSV Files."

This will launch the wizard.

In the first screen (Figure 5-16), it will ask you for the connection details to your SQL Server Master Instance which you can also choose from a drop-down list of active connections, if you're currently connected to the instance.

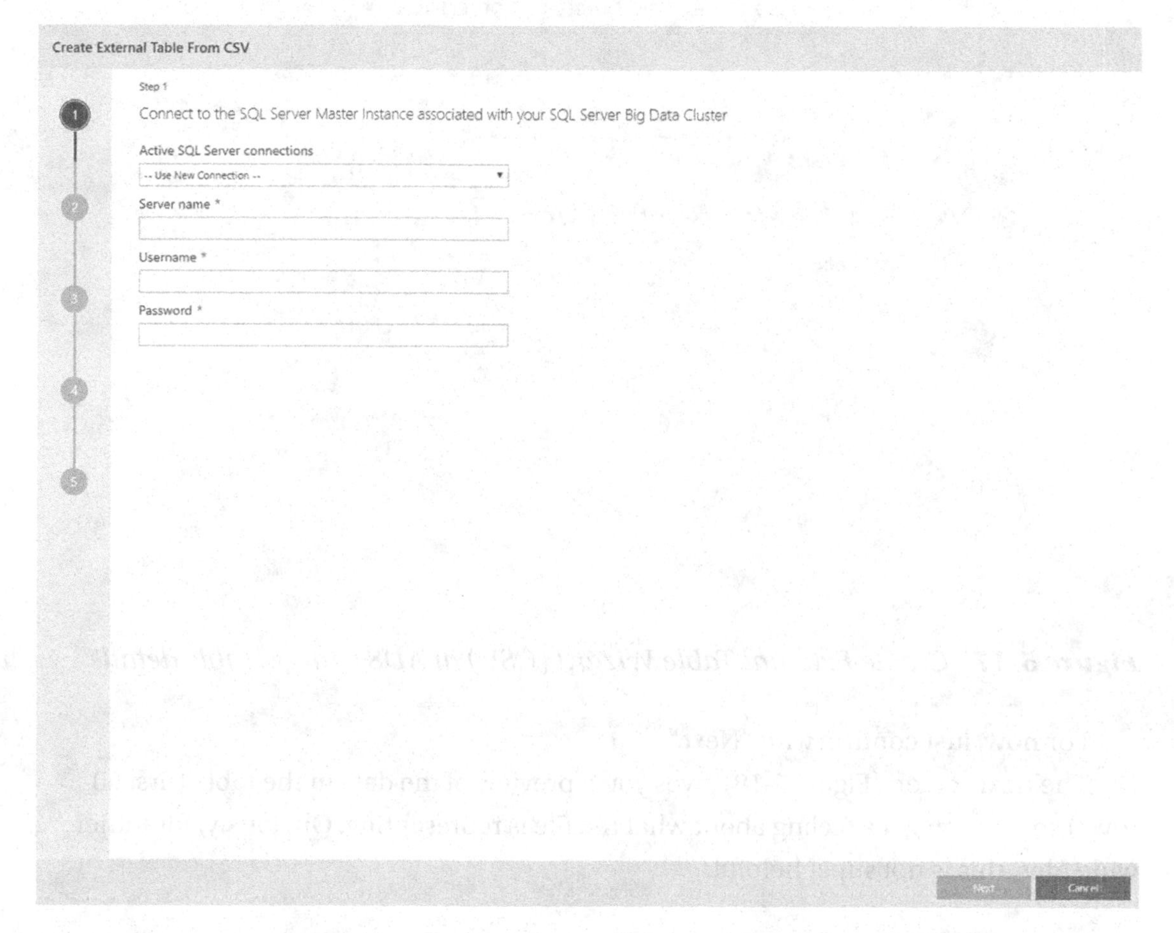


***Figure 5-16.*** *Create External Table Wizard (CSV) in ADS - select master instance*

Fill them in or select your connection and click "Next."

In the next step (Figure 5-17), the wizard will propose a target database as well as the name and the schema of the external table. All three can be modified if needed.

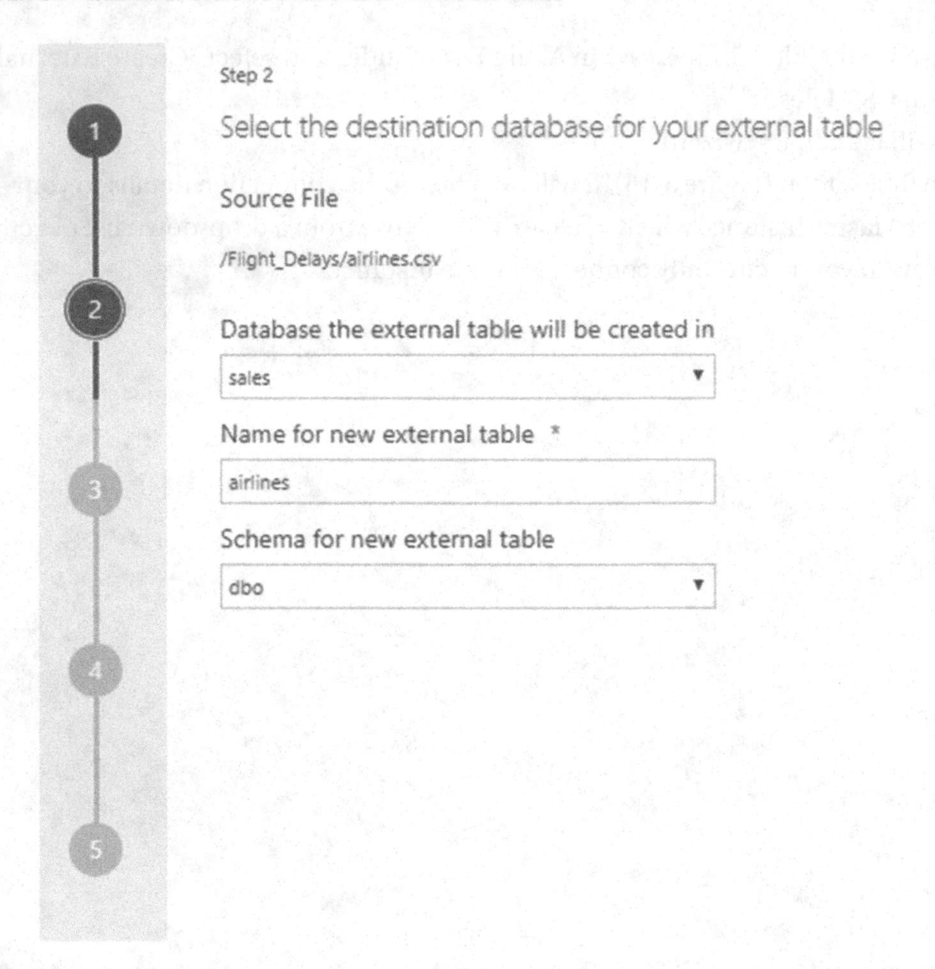


***Figure 5-17.*** *Create External Table Wizard (CSV) in ADS - target table details*

For now, just confirm with "Next."

The next screen (Figure 5-18) gives you a preview of the data in the table (first 50 rows) so you can get a feeling about what the file is representing. Obviously, for rather wide files, this is not super helpful.

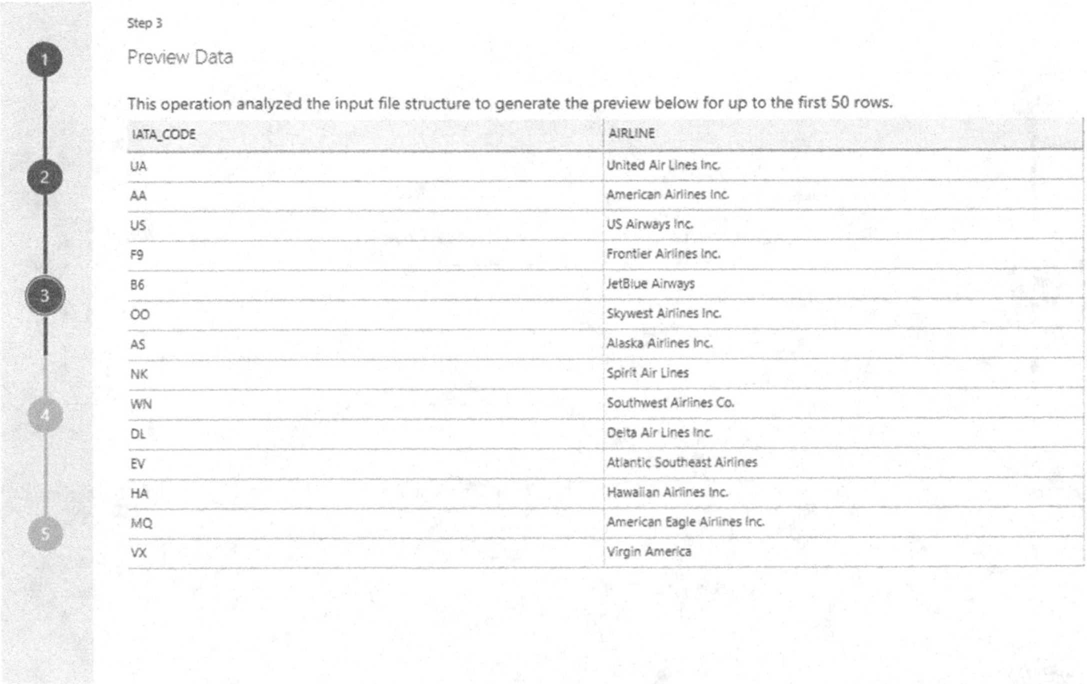

Step 3

Preview Data

This operation analyzed the input file structure to generate the preview below for up to the first 50 rows.

| IATA_CODE | AIRLINE |
|---|---|
| UA | United Air Lines Inc. |
| AA | American Airlines Inc. |
| US | US Airways Inc. |
| F9 | Frontier Airlines Inc. |
| B6 | JetBlue Airways |
| OO | Skywest Airlines Inc. |
| AS | Alaska Airlines Inc. |
| NK | Spirit Air Lines |
| WN | Southwest Airlines Co. |
| DL | Delta Air Lines Inc. |
| EV | Atlantic Southeast Airlines |
| HA | Hawaiian Airlines Inc. |
| MQ | American Eagle Airlines Inc. |
| VX | Virgin America |

***Figure 5-18.** Create External Table Wizard (CSV) in ADS - Preview Data*

There is nothing that can actually be done on this screen so just click "Next" again.

In step four (Figure 5-19), the wizard is proposing column names and data types. Both can be overridden. Unless you have a good reason to, in many cases it's actually a good advice to leave it unchanged as the detection mechanisms are rather solid so far.

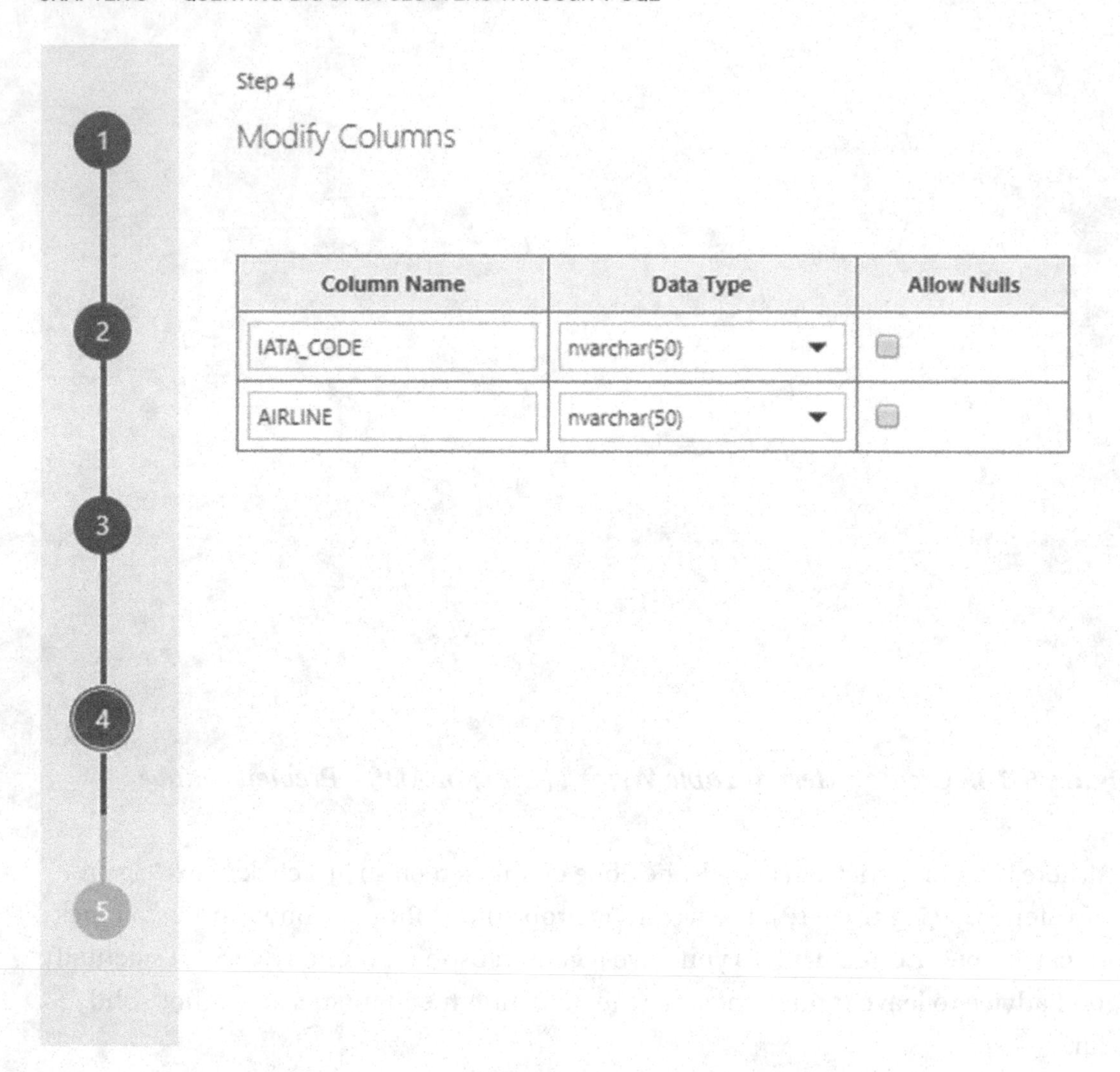


***Figure 5-19.*** *Create External Table Wizard (CSV) in ADS – Modify Columns*

After clicking "Next" again, we end up with a summary as shown in Figure 5-20 just like after the SQL Server table wizard.

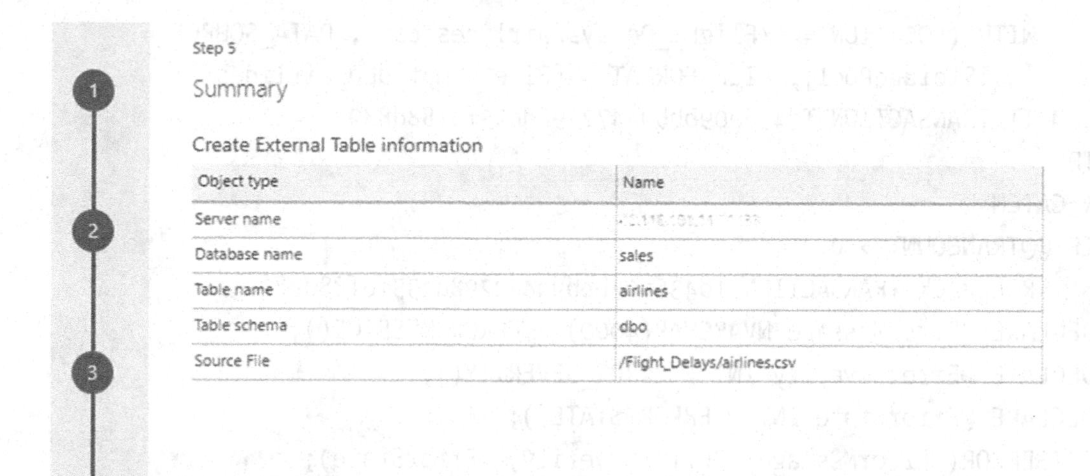


***Figure 5-20.*** *Create External Table Wizard (CSV) in ADS - Summary*

Choose "Generate Script" and click "Cancel." Take a look at the generated script (Listing 5-12).

***Listing 5-12.*** T-SQL output from the Create External Table Wizard (CSV) in ADS

```
BEGIN TRY
    BEGIN TRANSACTION Td436a09bbb9a472298de35f6f88d889
        USE [sales];
        CREATE EXTERNAL FILE FORMAT [FileFormat_dbo_airlines]
            WITH (FORMAT_TYPE = DELIMITEDTEXT, FORMAT_OPTIONS (FIELD_
            TERMINATOR = ',', STRING_DELIMITER = '"', FIRST_ROW = 2));
        CREATE EXTERNAL TABLE [dbo].[airlines]
        (
            [IATA_CODE] nvarchar(50) NOT NULL,
            [AIRLINE] nvarchar(50) NOT NULL
        )
```

```
        WITH (LOCATION = '/Flight_Delays/airlines.csv', DATA_SOURCE =
        [SqlStoragePool], FILE_FORMAT = [FileFormat_dbo_airlines]);
    COMMIT TRANSACTION Td436a09bbb9a472298de35f6f88d889
END TRY
BEGIN CATCH
    IF @@TRANCOUNT > 0
        ROLLBACK TRANSACTION Td436a09bbb9a472298de35f6f88d889
    DECLARE @ErrorMessage NVARCHAR(4000) = ERROR_MESSAGE();
    DECLARE @ErrorSeverity INT = ERROR_SEVERITY();
    DECLARE @ErrorState INT = ERROR_STATE();
    RAISERROR(@ErrorMessage, @ErrorSeverity, @ErrorState);
END CATCH;
```

As you can see, the wizard does not recycle identical file formats but rather creates one format per file. This has obviously pros and cons. The big pro is obviously that you won't end up with hundreds of formats. The big con is that you potentially start building dependencies between files where you don't want any. It is up to you whether you change the script to use the previously created csv_file format or just keep creating new formats for this exercise.

Repeat these steps for the other two files.

Then, we can query and join them (see Listing 5-13) as if they were SQL tables to get the ten Airline/Destination City combinations with the highest number of cancellations.

***Listing 5-13.*** SELECT statement against previously created external tables

```
SELECT TOP 10 ap.CITY, al.AIRLINE, COUNT(*)
FROM flights fl
    INNER JOIN airlines al
        ON fl.AIRLINE = al.IATA_CODE
    INNER JOIN airports ap
        ON fl.DESTINATION_AIRPORT = ap.IATA_CODE
WHERE cancelled = 1
GROUP BY ap.CITY,
         al.AIRLINE
ORDER BY COUNT(*) DESC;
```

This will result in an error.

The reason for that is the fact that the wizard only looks at the first 50 lines of your data, so if your data does not show a certain pattern in those rows, it won't be detected. The error message (Figure 5-21) clearly tells us what the issue is though.

```
Msg 7320, Level 16, State 110, Line 1
Cannot execute the query "Remote Query" against OLE DB provider "SQLNCLI11" for
linked server "(null)". 105082;Generic ODBC error: [Microsoft][ODBC Driver 17 for SQL Ser
ver][SQL Server]Bulk load data conversion error (overflow) for row 14500, column 27 (AIR_
SYSTEM_DELAY). .
```

***Figure 5-21.*** *Error message when querying the external tables*

In row 14500 (so way beyond the first 50), there is an overflow in the AIR_SYSTEM_DELAY column.

If we look at the script (Listing 5-14) for this table, we notice that this was detected as a tinyint.

***Listing 5-14.*** Original CREATE statement for external table flights

```
CREATE EXTERNAL TABLE [dbo].[flights]
      (
            [YEAR] smallint NOT NULL,
            [MONTH] tinyint NOT NULL,
            [DAY] tinyint NOT NULL,
            [DAY_OF_WEEK] tinyint NOT NULL,
            [AIRLINE] nvarchar(50) NOT NULL,
            [FLIGHT_NUMBER] smallint NOT NULL,
            [TAIL_NUMBER] nvarchar(50),
            [ORIGIN_AIRPORT] nvarchar(50) NOT NULL,
            [DESTINATION_AIRPORT] nvarchar(50) NOT NULL,
            [SCHEDULED_DEPARTURE] time NOT NULL,
            [DEPARTURE_TIME] time,
            [DEPARTURE_DELAY] smallint,
            [TAXI_OUT] tinyint,
            [WHEELS_OFF] time,
            [SCHEDULED_TIME] smallint NOT NULL,
```

```
        [ELAPSED_TIME] smallint,
        [AIR_TIME] smallint,
        [DISTANCE] smallint NOT NULL,
        [WHEELS_ON] time,
        [TAXI_IN] tinyint,
        [SCHEDULED_ARRIVAL] time NOT NULL,
        [ARRIVAL_TIME] time,
        [ARRIVAL_DELAY] smallint,
        [DIVERTED] bit NOT NULL,
        [CANCELLED] bit NOT NULL,
        [CANCELLATION_REASON] nvarchar(50),
        [AIR_SYSTEM_DELAY] tinyint,
        [SECURITY_DELAY] tinyint,
        [AIRLINE_DELAY] smallint,
        [LATE_AIRCRAFT_DELAY] smallint,
        [WEATHER_DELAY] tinyint
    )
    WITH (LOCATION = N'/Flight_Delays/flights.csv', DATA_SOURCE =
    [SqlStoragePool], FILE_FORMAT = [FileFormat_flights]);
```

Let's just change that to an int (or bigint) and, while we're at it, do the same for the other delay columns (Listing 5-15).

***Listing 5-15.*** Updated CREATE statement for external table flights

```
CREATE EXTERNAL TABLE [dbo].[flights]
    (
        [YEAR] smallint NOT NULL,
        [MONTH] tinyint NOT NULL,
        [DAY] tinyint NOT NULL,
        [DAY_OF_WEEK] tinyint NOT NULL,
        [AIRLINE] nvarchar(50) NOT NULL,
        [FLIGHT_NUMBER] smallint NOT NULL,
        [TAIL_NUMBER] nvarchar(50),
        [ORIGIN_AIRPORT] nvarchar(50) NOT NULL,
        [DESTINATION_AIRPORT] nvarchar(50) NOT NULL,
```

```
        [SCHEDULED_DEPARTURE] time NOT NULL,
        [DEPARTURE_TIME] time,
        [DEPARTURE_DELAY] smallint,
        [TAXI_OUT] tinyint,
        [WHEELS_OFF] time,
        [SCHEDULED_TIME] smallint NOT NULL,
        [ELAPSED_TIME] smallint,
        [AIR_TIME] smallint,
        [DISTANCE] smallint NOT NULL,
        [WHEELS_ON] time,
        [TAXI_IN] tinyint,
        [SCHEDULED_ARRIVAL] time NOT NULL,
        [ARRIVAL_TIME] time,
        [ARRIVAL_DELAY] smallint,
        [DIVERTED] bit NOT NULL,
        [CANCELLED] bit NOT NULL,
        [CANCELLATION_REASON] nvarchar(50),
        [AIR_SYSTEM_DELAY] bigint,
        [SECURITY_DELAY] bigint,
        [AIRLINE_DELAY] bigint,
        [LATE_AIRCRAFT_DELAY] bigint,
        [WEATHER_DELAY] bigint
    )
    WITH (LOCATION = N'/Flight_Delays/flights.csv', DATA_SOURCE =
    [SqlStoragePool], FILE_FORMAT = [FileFormat_flights]);
```

Note that you need to drop the external table (Listing 5-16) before you can re-create it.

***Listing 5-16.*** DROP statement

```
DROP EXTERNAL TABLE [dbo].[flights]
```

After dropping and re-creating the table, try running the cancellation query again. The error is gone! YAY!

Now try to get another view of that data: simply the first ten rows (Listing 5-17) with the matching columns from all three tables.

***Listing 5-17.*** Different SELECT statement against these external tables

```
SELECT TOP 10 *
FROM flights fl
    INNER JOIN airlines al
        ON fl.AIRLINE = al.IATA_CODE
    INNER JOIN airports ap
        ON fl.DESTINATION_AIRPORT = ap.IATA_CODE;
```

Unfortunately, this results in a new error (Figure 5-22).

```
Msg 7320, Level 16, State 110, Line 1
Cannot execute the query "Remote Query" against OLE DB provider "SQLNCLI11" for
linked server "(null)".
105081;Cannot load a null value into non-nullable target column.
```

***Figure 5-22.*** *Different error message when querying the external tables*

It's good practice to make all flat file columns nullable.

So, let's re-create our table (Listing 5-18) again, this time without any "NOT NULL" hints.

***Listing 5-18.*** DROP and CREATE external table flights again with all columns allowing NULLS

```
DROP EXTERNAL TABLE [dbo].[flights]
        GO
        CREATE EXTERNAL TABLE [dbo].[flights]
        (
            [YEAR] smallint,
            [MONTH] tinyint,
            [DAY] tinyint,
            [DAY_OF_WEEK] tinyint,
            [AIRLINE] nvarchar(50),
            [FLIGHT_NUMBER] smallint,
            [TAIL_NUMBER] nvarchar(50),
            [ORIGIN_AIRPORT] nvarchar(50),
```

```
        [DESTINATION_AIRPORT] nvarchar(50),
        [SCHEDULED_DEPARTURE] time,
        [DEPARTURE_TIME] time,
        [DEPARTURE_DELAY] smallint,
        [TAXI_OUT] tinyint,
        [WHEELS_OFF] time,
        [SCHEDULED_TIME] smallint,
        [ELAPSED_TIME] smallint,
        [AIR_TIME] smallint,
        [DISTANCE] smallint,
        [WHEELS_ON] time,
        [TAXI_IN] tinyint,
        [SCHEDULED_ARRIVAL] time,
        [ARRIVAL_TIME] time,
        [ARRIVAL_DELAY] smallint,
        [DIVERTED] bit,
        [CANCELLED] bit,
        [CANCELLATION_REASON] nvarchar(50),
        [AIR_SYSTEM_DELAY] bigint,
        [SECURITY_DELAY] bigint,
        [AIRLINE_DELAY] bigint,
        [LATE_AIRCRAFT_DELAY] bigint,
        [WEATHER_DELAY] bigint
    )
    WITH (LOCATION = N'/Flight_Delays/flights.csv', DATA_SOURCE =
    [SqlStoragePool], FILE_FORMAT = [FileFormat_flights]);
```

Now, both queries run smoothly! You will notice though that performance is not great which is mainly due to the lookup against the other two, rather small, flat files.

In such cases, we recommend to either store that data in a persisted or temporary SQL table (Listing 5-19).

***Listing 5-19.*** Store data from CSV in temp tables instead of direct queries

```
SELECT * into #al FROM airlines
SELECT * into #ap FROM airports
SELECT TOP 10 *
```

```
FROM flights fl
    INNER JOIN #al al
        ON fl.AIRLINE = al.IATA_CODE
    INNER JOIN #ap ap
        ON fl.DESTINATION_AIRPORT = ap.IATA_CODE;
DROP TABLE #al
DROP TABLE #ap
```

This improves performance tremendously without wasting too many system resources as the large dataset stays within the CSV.

One of the biggest challenges in such an environment is finding a good trade-off between data redundancy and performance.

## Accessing Data in an Azure Blob Storage

If you are storing data in an Azure Blob Storage, there is no need (apart from maybe network latency) to copy that data into your SqlStoragePool. You can also access a Blob Storage by defining it as an external data source (Listing 5-20).

***Listing 5-20.*** Create external data source using an Azure Blob Storage

```
CREATE EXTERNAL DATA SOURCE AzureStorage with (
      TYPE = HADOOP,
      LOCATION ='wasbs://<blob_container_name>@<azure_storage_account_
      name>.blob.core.windows.net',
      CREDENTIAL = AzureStorageCredential
);
```

# External Tables from Other Data Sources

## File-Based Data Sources

Other variations of file-based data sources are Parquet, Hive RCFile, and Hive ORC; however, if you are using the storage pool, only delimited and Parquet files are supported at this point.

As these are compressed file types, we need to provide the DATA_COMPRESSION as well as, for RCFile, the SERDE_METHOD (Listing 5-21).

***Listing 5-21.*** Sample CREATE statements for other external table types

```
-- Create an external file format for PARQUET files.
CREATE EXTERNAL FILE FORMAT file_format_name
WITH (
    FORMAT_TYPE = PARQUET
     [ , DATA_COMPRESSION = {
        'org.apache.hadoop.io.compress.SnappyCodec'
      | 'org.apache.hadoop.io.compress.GzipCodec'      }
    ]);

--Create an external file format for ORC files.
CREATE EXTERNAL FILE FORMAT file_format_name
WITH (
    FORMAT_TYPE = ORC
     [ , DATA_COMPRESSION = {
        'org.apache.hadoop.io.compress.SnappyCodec'
      | 'org.apache.hadoop.io.compress.DefaultCodec'      }
    ]);

--Create an external file format for RCFILE.
CREATE EXTERNAL FILE FORMAT file_format_name
WITH (
    FORMAT_TYPE = RCFILE,
    SERDE_METHOD = {
        'org.apache.hadoop.hive.serde2.columnar.LazyBinaryColumnarSerDe'
      | 'org.apache.hadoop.hive.serde2.columnar.ColumnarSerDe'
    }
    [ , DATA_COMPRESSION = 'org.apache.hadoop.io.compress.DefaultCodec' ]);
```

(taken from the official Microsoft Docs, `https://docs.microsoft.com/en-us/sql/t-sql/statements/create-external-file-format-transact-sql`)

# ODBC

When trying to connect to an external data source through a generic ODBC, you need to provide the server (and optionally a port) just like when adding a SQL Server, but in addition, you will need to specify the driver as well (Listing 5-22), so SQL Server knows how to connect to the source.

***Listing 5-22.*** Create external data source against an ODBC source

```
CREATE EXTERNAL DATA SOURCE <myODBCName>
WITH (
LOCATION = odbc://<ODBC server address>[:<port>],
CONNECTION_OPTIONS = 'Driver={<Name of Installed Driver>};
ServerNode = <name of server  address>:<Port>',
PUSHDOWN = ON,
 CREDENTIAL = credential_name
);
```

The PUSHDOWN flag defines whether computation will be pushed down to the source and is ON by default.

# Others

For the other built in, the syntax follows the generic schema (Listing 5-23).

***Listing 5-23.*** Generic CREATE statement for an external data source

```
CREATE EXTERNAL DATA SOURCE <myExternalDataSource>
WITH (
LOCATION = <vendor>://<server>[:<port>],
CREDENTIAL = credential_name
);
```

where you replace <vendor> by the name of the vendor of the built-in connector you're trying to use.

For Oracle, that would look like Listing 5-24.

***Listing 5-24.*** CREATE statement for an external data source in Oracle

```
CREATE EXTERNAL DATA SOURCE <myOracleSource>
WITH (
LOCATION = oracle://<server address>[:<port>],
CREDENTIAL = credential_name
```

The same logic applies for Teradata and MongoDB, and we expect all future vendors to be implemented in the same way.

# The SqlDataPool

The *SqlDataPool* is used to address the data pool of the Big Data Cluster which allows you to distribute data from a SQL table among the whole pool.

Just like with the storage pool, a respective pointer needs to be created first in each database that you want to use as shown in Listing 5-25.

***Listing 5-25.*** T-SQL code to create pointer to the SQL Data Pool

```
IF NOT EXISTS(SELECT * FROM sys.external_data_sources WHERE name = 'SqlDataPool')
    CREATE EXTERNAL DATA SOURCE SqlDataPool
    WITH (LOCATION = 'sqldatapool://controller-svc/default');
```

We will do this again in the sales database.

First of all, we start by creating an external table (Listing 5-26) again. Notice that unlike in previous samples, where we were addressing a specific file or location, in this instance, we only provide the hint that this table is to be stored in the SqlDataPool and is to be distributed using round robin.

***Listing 5-26.*** CREATE statement for external table on SqlDataPool

```
    CREATE EXTERNAL TABLE [web_clickstream_clicks_data_pool]
    ("wcs_user_sk" BIGINT , "i_category_id" BIGINT , "clicks" BIGINT)
    WITH
    (
        DATA_SOURCE = SqlDataPool,
        DISTRIBUTION = ROUND_ROBIN
    );
```

You can now insert data into this table using a regular INSERT INTO statement as shown in Listing 5-27.

***Listing 5-27.*** Populate table in SqlDataPool from SQL Query

```
INSERT INTO web_clickstream_clicks_data_pool
SELECT wcs_user_sk, i_category_id, COUNT_BIG(*) as clicks
  FROM sales.dbo.web_clickstreams_hdfs_parquet
 INNER JOIN sales.dbo.item it ON (wcs_item_sk = i_item_sk
                        AND wcs_user_sk IS NOT NULL)
 GROUP BY wcs_user_sk, i_category_id
```

Just as in all previous examples, this external table can now be queried and joined like any other table (Listing 5-28).

***Listing 5-28.*** SELECT against table in SqlDataPool (stand-alone and joined against other tables)

```
SELECT count(*) FROM [dbo].[web_clickstream_clicks_data_pool]
SELECT TOP 10 * FROM [dbo].[web_clickstream_clicks_data_pool]

SELECT TOP (100)
    w.wcs_user_sk,
    SUM( CASE WHEN i.i_category = 'Books' THEN 1 ELSE 0 END) AS book_
    category_clicks,
    SUM( CASE WHEN w.i_category_id = 1 THEN 1 ELSE 0 END) AS [Home & Kitchen],
    SUM( CASE WHEN w.i_category_id = 2 THEN 1 ELSE 0 END) AS [Music],
    SUM( CASE WHEN w.i_category_id = 3 THEN 1 ELSE 0 END) AS [Books],
    SUM( CASE WHEN w.i_category_id = 4 THEN 1 ELSE 0 END) AS [Clothing &
    Accessories],
    SUM( CASE WHEN w.i_category_id = 5 THEN 1 ELSE 0 END) AS [Electronics],
    SUM( CASE WHEN w.i_category_id = 6 THEN 1 ELSE 0 END) AS [Tools & Home
    Improvement],
    SUM( CASE WHEN w.i_category_id = 7 THEN 1 ELSE 0 END) AS [Toys & Games],
    SUM( CASE WHEN w.i_category_id = 8 THEN 1 ELSE 0 END) AS [Movies & TV],
    SUM( CASE WHEN w.i_category_id = 9 THEN 1 ELSE 0 END) AS [Sports & Outdoors]
  FROM [dbo].[web_clickstream_clicks_data_pool] as w
  INNER JOIN (SELECT DISTINCT i_category_id, i_category FROM item) as i
    ON i.i_category_id = w.i_category_id
GROUP BY w.wcs_user_sk;
```

The execution plan (Figure 5-23) looks similar to what we saw before when querying against the storage pool.

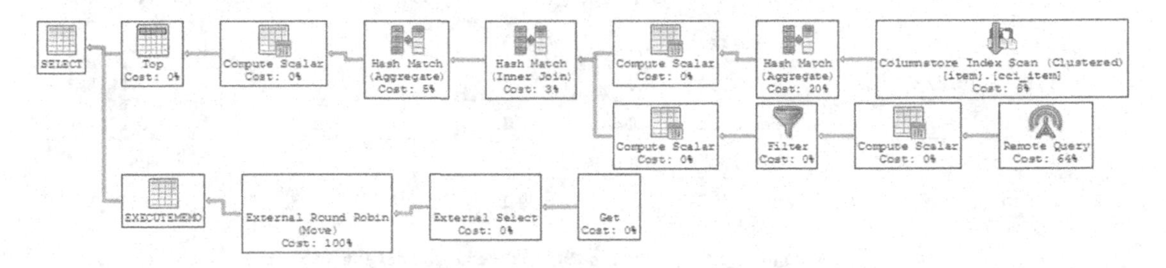


***Figure 5-23.*** *Execution plan in ADS*

Notice that unlike in the SqlStoragePool, the data that sits in the data pool is not visible as a file in the HDFS folder in Azure Data Studio, as its data is stored in regular SQL tables!

## Indexes on the SqlDataPool

One big difference between the SqlDataPool and other external data sources is that it consists of regular SQL Server tables, which means you can directly control indexes on those tables.

By default, every table in the SqlDataPool will be created with a clustered columnstore index. As the main use case for scale out and therefore the data pool are analytics workloads, this will probably satisfy many query needs out of the box.

If you still want to create or change an index, this needs to happen within the data pool itself, which can be achieved using the EXECUTE AT switch as shown in Listing 5-29.

***Listing 5-29.*** EXEC with EXECUTE AT

```
exec (<your query>) AT Data_Source SqlDataPool
```

You can use this to run any kind of query on the data pool and it will return one result grid for every node in the data pool. To get a list of all tables in the data pool, for example, run Listing 5-30.

***Listing 5-30.*** Get a list of tables in data pool

```
exec ('SELECT * FROM sys.tables') AT Data_Source SqlDataPool
```

The result is shown in Figure 5-24.

| | name | object_id | principal_id | schema_id | parent_object_id | type | type_desc |
|---|---|---|---|---|---|---|---|
| 1 | spt_fallback_db | 117575457 | NULL | 1 | 0 | U | USER_TABLE |
| 2 | spt_fallback_dev | 133575514 | NULL | 1 | 0 | U | USER_TABLE |
| 3 | spt_fallback_usg | 149575571 | NULL | 1 | 0 | U | USER_TABLE |
| 4 | spt_monitor | 1803153469 | NULL | 1 | 0 | U | USER_TABLE |
| 5 | MSreplication_options | 2107154552 | NULL | 1 | 0 | U | USER_TABLE |

| | name | object_id | principal_id | schema_id | parent_object_id | type | type_desc |
|---|---|---|---|---|---|---|---|
| 1 | spt_fallback_db | 117575457 | NULL | 1 | 0 | U | USER_TABLE |
| 2 | spt_fallback_dev | 133575514 | NULL | 1 | 0 | U | USER_TABLE |
| 3 | spt_fallback_usg | 149575571 | NULL | 1 | 0 | U | USER_TABLE |
| 4 | spt_monitor | 1803153469 | NULL | 1 | 0 | U | USER_TABLE |
| 5 | MSreplication_options | 2107154552 | NULL | 1 | 0 | U | USER_TABLE |

***Figure 5-24.*** *Result from a list of tables in data pool*

It may surprise you a little bit as you were maybe expecting different tables. The reason for that is that just like your master instance, the data pool will have multiple databases, so context is important. Let's try that again by switching the current database context as shown in Listing 5-31. This will only work for databases that have at least one table in SqlDataPool, as the database is created at that point.

***Listing 5-31.*** Get a list of tables in data pool from sales database

```
exec ('USE sales; SELECT * FROM sys.tables') AT Data_Source SqlDataPool
```

This time the result (see Figure 5-25) looks more as we would have expected.

Results Messages

| | name | object_id | principal_id | schema_id | parent_object_id | type | type_desc |
|---|---|---|---|---|---|---|---|
| 1 | web_clickstream_clicks_data_... | 581577110 | NULL | 1 | 0 | U | USER_TABLE |

| | name | object_id | principal_id | schema_id | parent_object_id | type | type_desc |
|---|---|---|---|---|---|---|---|
| 1 | web_clickstream_clicks_data_... | 581577110 | NULL | 1 | 0 | U | USER_TABLE |

***Figure 5-25.*** *Result from a list of tables in data pool from sales database*

To add an index, you would simply add the respective CREATE INDEX statement within the EXEC statement as illustrated in Listing 5-32.

***Listing 5-32.*** CREATE INDEX in data pool

```
EXEC ('USE Sales; CREATE NONCLUSTERED INDEX [CI_wcs_user_sk] ON [dbo].
[web_clickstream_clicks_data_pool] ([wcs_user_sk] ASC)') AT DATA_SOURCE
SqlDataPool
```

# Summary

In this chapter, we used the previously loaded data and queried it using multiple techniques using T-SQL, from another SQL Server to flat files. We also took a look on how certain tasks in that aspect can be automated.

As you've learned before, T-SQL isn't the only way to query a Big Data Cluster: another way of doing so is Spark. Chapter 6 will guide you through that process!

# CHAPTER 6

# Working with Spark in Big Data Clusters

So far, we have been querying data inside our SQL Server Big Data Cluster using external tables and T-SQL code. We do, however, have another method available to query data that is stored inside the HDFS filesystem of your Big Data Cluster. As you have read in Chapter 2, Big Data Clusters also have Spark included in the architecture, meaning we can leverage the power of Spark to query data stored inside our Big Data Cluster.

Spark is a very powerful option of analyzing and querying the data inside your Big Data Cluster, mostly because Spark is built as a distributed and parallel framework, meaning it is very fast at processing very large datasets making it far more efficient when you want to process large datasets than SQL Server. Spark also allows a large flexibility in terms of programming languages that it supports, the most prominent ones being Scala and PySpark (though Spark also supports R and Java).

The PySpark and Scala syntax are both very similar in the majority of commands we are using in the examples. There are some subtle nuances though.

The example code of Listings 6-1 and 6-2 shows how to read a CSV file into a data frame in both PySpark and Scala (don't worry we will get into more detail on data frames soon).

***Listing 6-1.*** Import CSV from HDFS using PySpark

```
# Import the airports.csv file from HDFS (PySpark)
df_airports = spark.read.format('csv').options(header='true',
inferSchema='true').load('/Flight_Delays/airports.csv')
```

***Listing 6-2.*** Import CSV from HDFS using Scala

```
// Import the airports.csv file from HDFS (Scala)
val df_airports = spark.read.format("csv").option("header", "true").
option("inferSchema", "true").load("/Flight_Delays/airports.csv")
```




B. Weissman and E. van de Laar, *SQL Server Big Data Clusters*,
https://doi.org/10.1007/978-1-4842-5985-6_6

As you can see, the code of the example looks very similar for PySpark and Scala, but there are some small differences. For instance, the character used for comments, in PySpark a comment is marked with a # sign, while in Scala we use //. Another difference is in the quotes. While we can use both a single quote and a double quote in the PySpark code, Scala is pickier accepting only double quotes. Also, where we don't need to specifically define a variable in PySpark (which is called a value in Scala), we do need to explicitly specify this when using Scala.

While this book is focused on Big Data Clusters, we believe an introduction to writing PySpark will be very useful when working with SQL Server Big Data Clusters since it allows you different method to work with the data inside your Big Data Cluster besides SQL.

# Loading Data and Creating a Spark Notebook

If you followed the steps in the "Getting Some Sample Files into the Installation" section of Chapter 4, you should have already imported the "2015 Flight Delays and Cancellations" dataset from Kaggle to the HDFS filesystem of your Big Data Cluster. If you haven't done so already, and want to follow along with the examples in this section, we recommend following the steps outlined in the "Getting Some Sample Files into the Installation" section before continuing. If you imported the dataset correctly, you should be able to see the "Flight_Delays" folder and the three CSV files inside the HDFS filesystem through Azure Data Studio as shown in Figure 6-1.

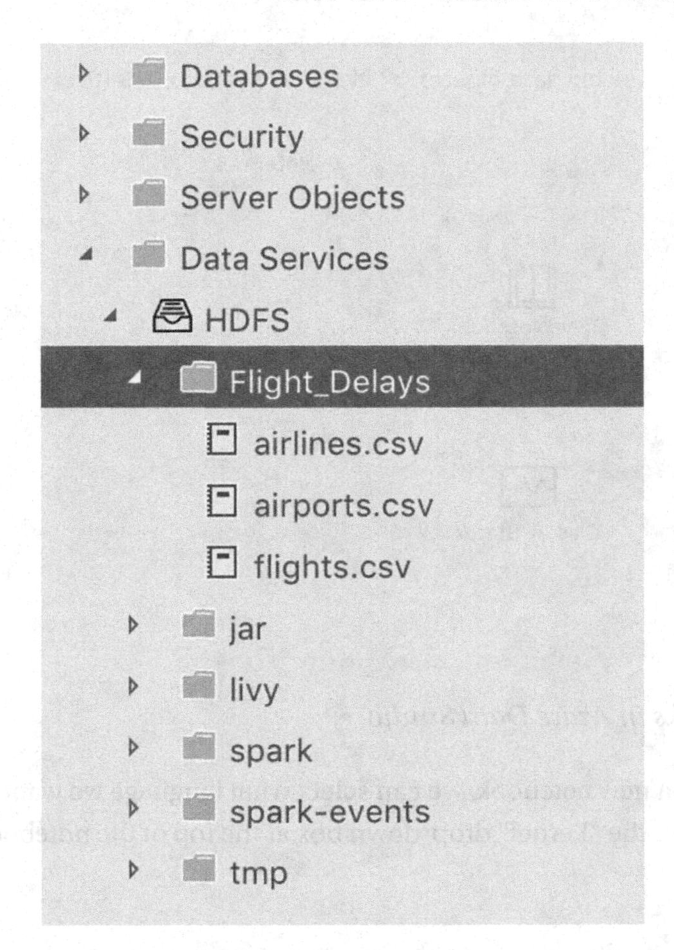


***Figure 6-1.*** *Flight delay files in HDFS store*

With our sample dataset available on HDFS, let's start with exploring the data a bit.

The first thing we need to do is to create a new notebook through the "New Notebook" option inside the Tasks window of our SQL Big Data Cluster tab (Figure 6-2).

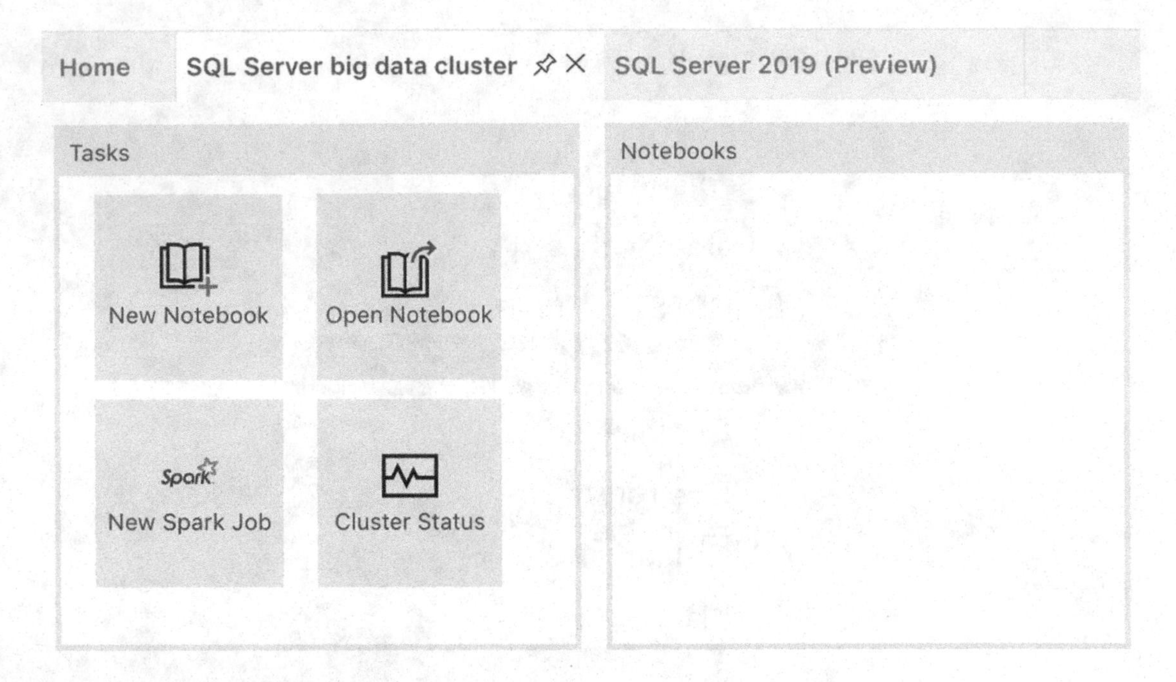


***Figure 6-2.** Tasks in Azure Data Studio*

After creating a new notebook, we can select what language we want to use by selecting it through the "Kernel" drop-down box at the top of the notebook as shown in Figure 6-3.

***Figure 6-3.** Kernel selection in ADS notebook*

During the remainder of this chapter, we will be using PySpark as the language of all the examples. If you want to follow along with the examples in this chapter, we recommend selecting the "PySpark" language.

With our notebook created and our language configured, let's look at our flight delay sample data!

## Working with Spark Data Frames

Now that we have access to our data, and we have a notebook, what we want to do is load our CSV data inside a data frame. Think of a data frame as a table-like structure that is created inside Spark. Conceptually speaking, a data frame is equal to a table inside SQL Server, but unlike a table that is generally stored on a single computer, a data frame consists of data distributed across (potentially) all the nodes inside your Spark cluster.

The code to load the data inside the "airports.csv" file into a Spark data frame can be seen in Listing 6-3. You can copy the code inside a cell of the notebook. All of the example code shown inside this chapter is best used as a separate cell inside a notebook. The full example notebook that contains all the code is available at this book's GitHub page.

***Listing 6-3.*** Import CSV data into data frame

```
# Import the airports.csv file from HDFS (PySpark)
df_airports = spark.read.format('csv').options(header='true',
inferSchema='true').load('/Flight_Delays/airports.csv')
```

If everything worked, you should end up with a Spark data frame that contains the data from the airports.csv file.

As you can see from the example code, we provided a number of options to the `spark.read.format` command. The most important one is the type of file we are reading; in our case this is a CSV file. The other options we provide tell Spark how to handle the CSV file. By setting the option `header='true'`, we specify that the CSV file has a header row which contains the column names. The option `inferSchema='true'` helps us with automatically detecting what datatypes we are dealing with in each column. If we do not specify this option, or set it to false instead, all the columns will be set to a string datatype instead of the datatype we would expect our data to be (e.g., an integer datatype for numerical data). If you do not use the `inferSchema='true'` option, or inferSchema configures the wrong datatypes, you have to define the schema before importing the CSV file and supply it as an option to the `spark.read.format` command as shown in the example code of Listing 6-4.

***Listing 6-4.*** Define schema and supply it to spark.read.format

```
# Manually set the schema and supply it to the spark.read function

# We need to import the pyspark.sql.types library
from pyspark.sql.types import *

df_schema = StructType([
            StructField("IATA_CODE", StringType(), True),
            StructField("AIRPORT", StringType(), True),
            StructField("CITY", StringType(), True),
            StructField("STATE", StringType(), True),
            StructField("COUNTRY", StringType(), True),
            StructField("LATITUDE", DoubleType(), True),
            StructField("LONGITUDE", DoubleType(), True)
            ])

# With the schema declared, we can supply it to the spark.read function
df_airports = spark.read.format('csv').options(header='true').schema(df_
schema).load('/Flight_Delays/airports.csv')
```

If this was your first notebook command against the Spark cluster, you will get some information back regarding the creation of a Spark session as you can see in Figure 6-4.

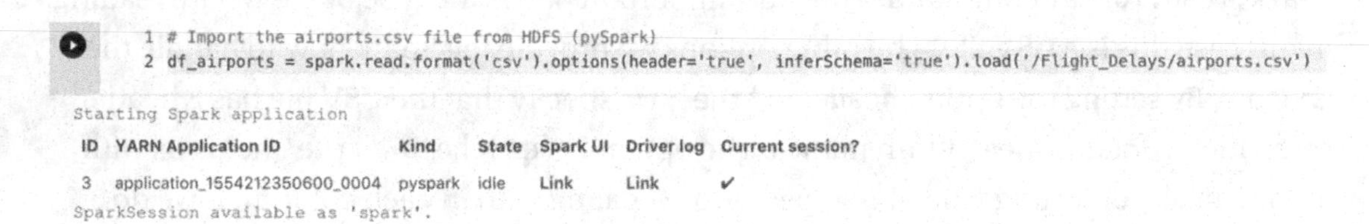


***Figure 6-4.*** *Output of spark.read.format*

A Spark session represents your entry point to interact with the Spark functions. In the past, we would have to define a Spark context to connect to the Spark cluster, and depending on what functionality we needed, we would have to create a separate context for that specific functionality (like Spark SQL, or streaming functionalities). Starting from Spark 2.0, Spark sessions became available as entry point, and it, by default, includes all kinds of different functions that we had to create a separate context for in the past, making it easier to work with them. When we run the first command inside a notebook against the Spark cluster, a Spark session needs to be created so we are able to send commands to the cluster. Running subsequent commands will make use of the initially created Spark session.

Now that we have our CSV data inside a data frame we can run all kind of commands to retrieve information about our data frame. For instance, the example of Listing 6-5 returns the number of rows that are inside the data frame.

***Listing 6-5.*** Retrieve row count of the data frame

```
# Display the amount of rows inside the df_airports data frame
df_airports.count()
```

The result should be 322 rows as shown in Figure 6-5 which returns the number of rows.

```
[4]   1 # Display the amount of rows inside the df_airports dataframe
      2 df_airports.count()

322
```

***Figure 6-5.*** *Output of row count*

Another very useful command is to return the schema of the data frame. This shows us which columns make up the data frame and their datatypes. The code of Listing 6-6 gets the schema of the `df_airports` data frame and returns it as output (Figure 6-6).

***Listing 6-6.*** Retrieve the schema of the data frame

```
# Display the schema of the df_flights data frame (PySpark)
df_airports.printSchema()
```

```
root
 |-- IATA_CODE: string (nullable = true)
 |-- AIRPORT: string (nullable = true)
 |-- CITY: string (nullable = true)
 |-- STATE: string (nullable = true)
 |-- COUNTRY: string (nullable = true)
 |-- LATITUDE: double (nullable = true)
 |-- LONGITUDE: double (nullable = true)
```

***Figure 6-6.*** *Schema output of the df_airports data frame*

Now that we know the schema of the data frame, let's return some data from it. One easy way to do that is to use the `head()` function. This function will return the top *n* rows from the data frame. In the following example, we will return the first row of the data frame. The output of the command is shown below the example (Listing 6-7 is followed by Figure 6-7).

***Listing 6-7.*** Retrieve first row of data frame

```
# Let's return the first row
df_airports.head(1)
```

```
[Row(IATA_CODE=u'ABE', AIRPORT=u'Lehigh Valley International Airport', CITY=u'Allentown', STATE=u'PA', COUNTRY=u'USA'
```

***Figure 6-7.*** *First row of the df_airports data frame*

As you can see in Figure 6-7, the results aren't by default returned in a table-like structure. This is because `head()` only returns the output as a string-like structure.

To return a table structure when getting data from a dataset, you can use the `show()` function as shown in the following example. Show() accepts an integer as a parameter to indicate how many rows should be returned. In the example (Listing 6-8), we supplied a value of 10, indicating we want the top ten rows returned by the function (Figure 6-8).

***Listing 6-8.*** Retrieve first row of data frame as a table structure

```
# Select top ten rows, return as a table structure
df_airports.show(10)
```

```
+---------+--------------------+-------------+-----+-------+--------+----------+
|IATA_CODE|             AIRPORT|         CITY|STATE|COUNTRY|LATITUDE| LONGITUDE|
+---------+--------------------+-------------+-----+-------+--------+----------+
|      ABE|Lehigh Valley Int...|    Allentown|   PA|    USA|40.65236|  -75.4404|
|      ABI|Abilene Regional ...|      Abilene|   TX|    USA|32.41132|  -99.6819|
|      ABQ|Albuquerque Inter...|  Albuquerque|   NM|    USA|35.04022|-106.60919|
|      ABR|Aberdeen Regional...|     Aberdeen|   SD|    USA|45.44906| -98.42183|
|      ABY|Southwest Georgia...|       Albany|   GA|    USA|31.53552| -84.19447|
|      ACK|Nantucket Memoria...|    Nantucket|   MA|    USA|41.25305| -70.06018|
|      ACT|Waco Regional Air...|         Waco|   TX|    USA|31.61129| -97.23052|
|      ACV|      Arcata Airport|Arcata/Eureka|   CA|    USA|40.97812|-124.10862|
|      ACY|Atlantic City Int...|Atlantic City|   NJ|    USA|39.45758| -74.57717|
|      ADK|        Adak Airport|         Adak|   AK|    USA|51.87796|-176.64603|
+---------+--------------------+-------------+-----+-------+--------+----------+
only showing top 10 rows
```

***Figure 6-8.*** *Show the top ten rows of the df_airports data frame*

Next to returning the entire data frame, we can, just like in SQL, select a subset of the data based on the columns we are interested in. The following example (Listing 6-9) only returns the top ten rows of the AIRPORT and CITY columns of the `df_airports` data frame (Figure 6-9).

***Listing 6-9.*** Select specific columns of the first ten rows of the data frame

```
# We can also select specific columns from the data frame
df_airports.select('AIRPORT','CITY').show(10)
```

```
+--------------------+--------------+
|             AIRPORT|          CITY|
+--------------------+--------------+
|Lehigh Valley Int...|     Allentown|
|Abilene Regional ...|       Abilene|
|Albuquerque Inter...|   Albuquerque|
|Aberdeen Regional...|      Aberdeen|
|Southwest Georgia...|        Albany|
|Nantucket Memoria...|     Nantucket|
|Waco Regional Air...|          Waco|
|      Arcata Airport|Arcata/Eureka|
|Atlantic City Int...|Atlantic City|
|        Adak Airport|          Adak|
+--------------------+--------------+
only showing top 10 rows
```

***Figure 6-9.*** *Top ten rows for the AIRPORT and CITY column of the df_airports data frame*

Just like in SQL, we also have the ability to sort the data based on one or multiple columns. In the following example (Listing 6-10), we are retrieving the top ten rows of the `df_airports` data frame order first by STATE descending, then by CITY descending (Figure 6-10).

***Listing 6-10.*** Retrieve the first ten rows of the data frame using sorting

```
# We can also sort on one or multiple columns
df_airports.orderBy(df_airports.STATE.desc(), df_airports.CITY.desc()).show(10)
```

```
+---------+--------------------+-------------+-----+-------+--------+----------+
|IATA_CODE|             AIRPORT|         CITY|STATE|COUNTRY|LATITUDE| LONGITUDE|
+---------+--------------------+-------------+-----+-------+--------+----------+
|      COD|Yellowstone Regio...|         Cody|   WY|    USA|44.52019| -109.0238|
|      RKS|Rock Springs-Swee...|Rock Springs|   WY|    USA|41.59422|-109.06519|
|      GCC|Gillette-Campbell...|     Gillette|   WY|    USA| 44.3489|-105.53936|
|      JAC|Jackson Hole Airport|      Jackson|   WY|    USA|43.60732|-110.73774|
|      CPR|Natrona County In...|       Casper|   WY|    USA|42.90836|-106.46447|
|      LAR|Laramie Regional ...|      Laramie|   WY|    USA|41.31205|-105.67499|
|      CRW|      Yeager Airport|   Charleston|   WV|    USA|38.37315| -81.59319|
|      GRB|Green Bay-Austin ...|    Green Bay|   WI|    USA|44.48507| -88.12959|
|      CWA|Central Wisconsin...|      Mosinee|   WI|    USA|44.77762| -89.66678|
|      EAU|Chippewa Valley R...|   Eau Claire|   WI|    USA|44.86526| -91.48507|
+---------+--------------------+-------------+-----+-------+--------+----------+
only showing top 10 rows
```

***Figure 6-10.*** *df_airports data frame sorted on STATE and CITY columns*

So far when getting data from the data frame, we have been selecting the top *n* rows but we can also filter the data frame on a specific column value. To do this, we can add the `filter` function and supply a predicate. The following example (Listing 6-11) filters the data frame and only returns information about airports located in the city Jackson (Figure 6-11).

***Listing 6-11.*** Filter a date frame

```
# Filter results based on a column value
df_airports.filter(df_airports.CITY == 'Jackson').show()
```

```
+---------+--------------------+-------+-----+-------+--------+----------+
|IATA_CODE|             AIRPORT|   CITY|STATE|COUNTRY|LATITUDE| LONGITUDE|
+---------+--------------------+-------+-----+-------+--------+----------+
|      JAC|Jackson Hole Airport|Jackson|   WY|    USA|43.60732|-110.73774|
|      JAN|Jackson-Evers Int...|Jackson|   MS|    USA|32.31117| -90.07589|
+---------+--------------------+-------+-----+-------+--------+----------+
```

***Figure 6-11.*** *df_airports filtered on the CITY column*

Besides filtering on a single value, we can also supply a list of values we want to filter on, much in the same way as you would use the IN clause in SQL. The code shown in Listing 6-12 results in Figure 6-12.

***Listing 6-12.*** Multifiltering a data frame

```
# Besides filtering on a single value, we can also use IN to supply
multiple filter items

# We need to import the col function from pyspark.sql.functions
from pyspark.sql.functions import col

# Declare a list with city names
city_list = ['Jackson','Charleston','Cody']

# Filter the data frame
df_airports.where(col("CITY").isin(city_list)).show()
```

```
+---------+--------------------+----------+-----+-------+--------+----------+
|IATA_CODE|             AIRPORT|      CITY|STATE|COUNTRY|LATITUDE| LONGITUDE|
+---------+--------------------+----------+-----+-------+--------+----------+
|      CHS|Charleston Intern...|Charleston|   SC|    USA|32.89865| -80.04051|
|      COD|Yellowstone Regio...|      Cody|   WY|    USA|44.52019| -109.0238|
|      CRW|      Yeager Airport|Charleston|   WV|    USA|38.37315| -81.59319|
|      JAC|Jackson Hole Airport|   Jackson|   WY|    USA|43.60732|-110.73774|
|      JAN|Jackson-Evers Int...|   Jackson|   MS|    USA|32.31117| -90.07589|
+---------+--------------------+----------+-----+-------+--------+----------+
```

***Figure 6-12.*** *Filtering on multiple values stored in a list*

In the example in Figure 6-12, we declared a list of values we want to filter on and used the `isin` function to supply the list to the where function.

Besides the == operator to indicate values should be equal, there are multiple operators available when filtering data frames which you are probably already familiar with, the most frequently used are shown in Table 6-1.

***Table 6-1.*** *Compare operators in PySpark*

| | |
|---|---|
| == | Equal to |
| | Unequal to |
| < | Less than |
| <= | Less than or equal |
| > | Greater than |
| >= | Greater than or equal |

We have been focusing on getting data out of the data frame so far. However, there might also be situations where you want to remove a row from the data frame, or perhaps, update a specific value. Generally speaking, updating values inside Spark data frames is not as straightforward as, for instance, writing an UPDATE statement in SQL, which updates the value in the actual table. In most situations, updating rows inside data frames revolves around creating a sort of mapping data frame and joining your original data frame to the mapping data frame and storing that as a new data frame. This way your final data frame contains the updates.

Simply speaking, you perform a selection on the data you want to update (as shown in Figure 6-13) and return the updated value for the row you want to update and save that to a new data frame as shown in the example code of Listing 6-13.

***Listing 6-13.*** Perform multiple actions on the data frame

```
# Update a row
# We need to import the col and when function from pyspark.sql.functions
from pyspark.sql.functions import col, when

# Select the entire data frame but set the CITY value to "Cody" instead of
"Jackson" where the IATA_CODE = "COD"
# Store the results in the new df_airports_updated data frame
df_airports_updated = df_airports.withColumn("CITY", when(col("IATA_CODE") ==
"COD", "Jackson"))

# Return the results, filter on IATA_CODE == "COD"
df_airports_updated = df_airports_updated.filter(df_airports_updated.IATA_
CODE == 'COD').show()
```

```
+---------+--------------------+-------+-----+-------+--------+---------+
|IATA_CODE|             AIRPORT|   CITY|STATE|COUNTRY|LATITUDE|LONGITUDE|
+---------+--------------------+-------+-----+-------+--------+---------+
|      COD|Yellowstone Regio...|Jackson|   WY|    USA|44.52019|-109.0238|
+---------+--------------------+-------+-----+-------+--------+---------+
```

***Figure 6-13.*** *Updated value to Jackson instead of Cody*

If we are interested in removing rows the same reasoning applies, we do not physically delete rows from the data frame; instead, we perform a selection that does not include the rows we want removed and store that as a separate data frame (Figure 6-14 shows the result of using Listing 6-14).

***Listing 6-14.*** Remove a row from a data frame

```
# Remove a row

# Select the entire data frame except where the IATA_CODE = "COD"
# Store the results in the new df_airports_removed data frame
df_airports_removed = df_airports.filter(df_airports.IATA_CODE <> "COD")

# Return the results, filter on IATA_CODE == "COD"
df_airports_removed.filter(df_airports_removed.IATA_CODE == "COD").show()
```

```
+---------+-------+----+-----+-------+--------+---------+
|IATA_CODE|AIRPORT|CITY|STATE|COUNTRY|LATITUDE|LONGITUDE|
+---------+-------+----+-----+-------+--------+---------+
+---------+-------+----+-----+-------+--------+---------+
```

***Figure 6-14.*** *No data is present where IATA_CODE == "COD"*

The concept of no deleting or updating the physical data itself but rather working through selections to perform update or delete operations (and then store them in new data frames) is very important when working with data frames inside Spark and has everything to do with the fact that a data frame is only a logical representation of the data stored inside your Spark cluster. A data frame behaves more like a view to the data stored in one or multiple files inside your Spark cluster. While we can filter and modify the way the view returns the data, we cannot modify the data through the view itself.

One final example we want to show before we continue with a number of more advanced data frame processing examples is grouping data. Grouping data based on columns in a data frame is very useful when you want to perform aggregations or calculations based on the data inside the data frame column. For instance, in the following example code (Listing 6-15), we perform a count of how many airports a distinct city has in the `df_airports` data frame (Figure 6-15).

***Listing 6-15.*** Group a data frame

```
# Count the number of airports of each city and sort on the count descending
df_airports.groupby("City").count().sort(col("count").desc()).show(10)
```

```
+-----------+-----+
|       City|count|
+-----------+-----+
|  Rochester|    2|
| Wilmington|    2|
|    Jackson|    2|
|     Albany|    2|
|    Chicago|    2|
| Charleston|    2|
|Springfield|    2|
|  San Diego|    2|
|   Portland|    2|
|   Columbus|    2|
+-----------+-----+
only showing top 10 rows
```

***Figure 6-15.*** *Count the number of airports for each unique city*

In this example, we used the `sort` function instead of the `orderBy` we used earlier in this section to sort the results. Both functions are essentially identical (orderBy is actually an alias for sort) and there is no difference in terms of functionality between both functions.

## More Advanced Data Frame Handling

So far, we've looked at – relatively – simple operations we can perform on data frames like selecting specific rows, grouping data, and ordering data frames.

However, there are far more things we can do in Spark when it comes to data frames, for instance, joining multiple data frames together into a new data frame. In this section, we are going to focus on doing some more advanced data frame wrangling.

To start off, so far, we have been working with a single dataset which we imported into a data frame that contains information of the various airports in America. In many situations you do not have a single dataset that contains everything you need, meaning you will end up with multiple data frames. Using PySpark we can join these data frames together on a key the data frames share and build a new, joined, data frame.

Before we can get started on joining data frames together, we will need to import the other sample datasets from the 2015 Flight Delays and Cancellations examples we are working with. If you are following along with the examples in this chapter, you should already have a data frame called `df_airports` that contains the data of the airports.csv file. If you haven't, you can run the following code (Listing 6-16) to import the data from the file into a data frame.

***Listing 6-16.*** Import airports.csv into data frame

```
df_airports = spark.read.format('csv').options(header='true',
inferSchema='true').load('/Flight_Delays/airports.csv')
```

We can use the same command (Listing 6-17) to import the other two CSV files: airlines.csv and flights.csv.

***Listing 6-17.*** Import airlines.csv and flights.csv into data frames

```
# Importing the other CSV files into data frames as well
df_airlines = spark.read.format('csv').options(header='true',
inferSchema='true').load('/Flight_Delays/airlines.csv')
df_flights = spark.read.format('csv').options(header='true',
inferSchema='true').load('/Flight_Delays/flights.csv')
```

After executing Listing 6-17, we should have three separate data frames available to us in the PySpark notebook: `df_airports`, `df_airlines`, and `df_flights`.

To join two data frames, we have to supply the key on which we are joining the two data frames on. If this key is identical on both data frames, we do not have to explicitly set the mapping in the join (and we only need to supply the column name as a parameter). However, we believe it is always good practice to describe the join mapping to make the code easier to understand. Also, in the sample dataset we are using, the data frame columns have different column names on which we need to join requiring an explicit join mapping.

The code example of Listing 6-18 will join the `df_flights` and `df_airlines` data frames together using an inner join and output a new data frame called `df_flightinfo`. We return the schema of the newly created data frame to see how the two data frames are joined together (Figure 6-16).

***Listing 6-18.*** Join two data frames and retrieve the schema of the result

```
from pyspark.sql.functions import *

# Let's join the df_airlines and df_flights data frames using an inner join
on the airline code
df_flightinfo = df_flights.join(df_airlines, df_flights.AIRLINE == df_
airlines.IATA_CODE, how="inner")

# Print the schema of the joined data frame
df_flightinfo.printSchema()
```

```
root
 |-- YEAR: integer (nullable = true)
 |-- MONTH: integer (nullable = true)
 |-- DAY: integer (nullable = true)
 |-- DAY_OF_WEEK: integer (nullable = true)
 |-- AIRLINE: string (nullable = true)
 |-- FLIGHT_NUMBER: integer (nullable = true)
 |-- TAIL_NUMBER: string (nullable = true)
 |-- ORIGIN_AIRPORT: string (nullable = true)
 |-- DESTINATION_AIRPORT: string (nullable = true)
 |-- SCHEDULED_DEPARTURE: integer (nullable = true)
 |-- DEPARTURE_TIME: integer (nullable = true)
 |-- DEPARTURE_DELAY: integer (nullable = true)
 |-- TAXI_OUT: integer (nullable = true)
 |-- WHEELS_OFF: integer (nullable = true)
 |-- SCHEDULED_TIME: integer (nullable = true)
 |-- ELAPSED_TIME: integer (nullable = true)
 |-- AIR_TIME: integer (nullable = true)
 |-- DISTANCE: integer (nullable = true)
 |-- WHEELS_ON: integer (nullable = true)
 |-- TAXI_IN: integer (nullable = true)
 |-- SCHEDULED_ARRIVAL: integer (nullable = true)
 |-- ARRIVAL_TIME: integer (nullable = true)
 |-- ARRIVAL_DELAY: integer (nullable = true)
 |-- DIVERTED: integer (nullable = true)
 |-- CANCELLED: integer (nullable = true)
 |-- CANCELLATION_REASON: string (nullable = true)
 |-- AIR_SYSTEM_DELAY: integer (nullable = true)
 |-- SECURITY_DELAY: integer (nullable = true)
 |-- AIRLINE_DELAY: integer (nullable = true)
 |-- LATE_AIRCRAFT_DELAY: integer (nullable = true)
 |-- WEATHER_DELAY: integer (nullable = true)
 |-- IATA_CODE: string (nullable = true)
 |-- AIRLINE: string (nullable = true)
```

***Figure 6-16.*** *Schema of the df_flightinfo data frame which is a join between df_flights and df_airlines*

As you can see in Figure 6-16, the two columns (IATA_CODE and AIRLINE) that make up the `df_airlines` data frame are added to the right side of the new `df_flightinfo` data frame. Because we already have the IATA_CODE in the `df_flights` data frame, we end up having duplicate columns in the new data frame (to make matters more interesting: in this sample dataset the `df_flights` data frame uses the column "AIRLINE" to denote the IATA code on which we join the `df_airlines` data frame. The `df_airlines` data frame also has the AIRLINE column but it shows the full airline name. Essentially this means both AIRLINE columns in the `df_flightinfo` data frame contain different data).

We can easily drop the duplicate column when joining both data frames by specifying it in the join command (Listing 6-19 and Figure 6-17).

***Listing 6-19.*** Joining two data frames while dropping a column

```
from pyspark.sql.functions import *

# We will join both data frames again but this time drop the AIRLINE column
of the df_flights data frame
df_flightinfo = df_flights.join(df_airlines, df_flights.AIRLINE == df_
airlines.IATA_CODE, how="inner").drop(df_flights.AIRLINE)

# Print the schema of the joined data frame
df_flightinfo.printSchema()
```

```
root
 |-- YEAR: integer (nullable = true)
 |-- MONTH: integer (nullable = true)
 |-- DAY: integer (nullable = true)
 |-- DAY_OF_WEEK: integer (nullable = true)
 |-- FLIGHT_NUMBER: integer (nullable = true)
 |-- TAIL_NUMBER: string (nullable = true)
 |-- ORIGIN_AIRPORT: string (nullable = true)
 |-- DESTINATION_AIRPORT: string (nullable = true)
 |-- SCHEDULED_DEPARTURE: integer (nullable = true)
 |-- DEPARTURE_TIME: integer (nullable = true)
 |-- DEPARTURE_DELAY: integer (nullable = true)
 |-- TAXI_OUT: integer (nullable = true)
 |-- WHEELS_OFF: integer (nullable = true)
 |-- SCHEDULED_TIME: integer (nullable = true)
 |-- ELAPSED_TIME: integer (nullable = true)
 |-- AIR_TIME: integer (nullable = true)
 |-- DISTANCE: integer (nullable = true)
 |-- WHEELS_ON: integer (nullable = true)
 |-- TAXI_IN: integer (nullable = true)
 |-- SCHEDULED_ARRIVAL: integer (nullable = true)
 |-- ARRIVAL_TIME: integer (nullable = true)
 |-- ARRIVAL_DELAY: integer (nullable = true)
 |-- DIVERTED: integer (nullable = true)
 |-- CANCELLED: integer (nullable = true)
 |-- CANCELLATION_REASON: string (nullable = true)
 |-- AIR_SYSTEM_DELAY: integer (nullable = true)
 |-- SECURITY_DELAY: integer (nullable = true)
 |-- AIRLINE_DELAY: integer (nullable = true)
 |-- LATE_AIRCRAFT_DELAY: integer (nullable = true)
 |-- WEATHER_DELAY: integer (nullable = true)
 |-- IATA_CODE: string (nullable = true)
 |-- AIRLINE: string (nullable = true)
```

***Figure 6-17.** df_flightinfo schema without the duplicate AIRLINE column*

As you can see from the schema shown in Figure 6-17, we now end up with only one AIRLINE column which contains the data we are after (the full airline name).

With the duplicate column removed, let's select some information from the new df_flightinfo data frame. For this example, let's say we are interested in seeing the scheduled and actual elapsed travel times for each flight together with the airline that performed the flight. We can simply select the columns we are interested in just as we did a number of times already in this chapter. This time using the code shown in Listing 6-20 results in the table shown in Figure 6-18.

***Listing 6-20.*** Select a number of columns from the joined data frame

```
# Select a number of columns from the joined data frame
df_flightinfo.select("FLIGHT_NUMBER", "AIRLINE", "SCHEDULED_TIME",
"ELAPSED_TIME").show()
```

```
+-------------+--------------------+--------------+------------+
|FLIGHT_NUMBER|             AIRLINE|SCHEDULED_TIME|ELAPSED_TIME|
+-------------+--------------------+--------------+------------+
|           98|Alaska Airlines Inc.|           205|         194|
|         2336|American Airlines...|           280|         279|
|          840|     US Airways Inc.|           286|         293|
|          258|American Airlines...|           285|         281|
|          135|Alaska Airlines Inc.|           235|         215|
|          806|Delta Air Lines Inc.|           217|         230|
|          612|    Spirit Air Lines|           181|         170|
|         2013|     US Airways Inc.|           273|         249|
|         1112|American Airlines...|           195|         193|
|         1173|Delta Air Lines Inc.|           221|         203|
|         2336|Delta Air Lines Inc.|           173|         149|
|         1674|American Airlines...|           268|         266|
|         1434|Delta Air Lines Inc.|           214|         210|
|         2324|Delta Air Lines Inc.|           215|         199|
|         2440|Delta Air Lines Inc.|           189|         198|
|          108|Alaska Airlines Inc.|           204|         194|
|         1560|Delta Air Lines Inc.|           210|         200|
|         1197|United Air Lines ...|           218|         217|
|          122|Alaska Airlines Inc.|           215|         201|
|         1670|Delta Air Lines Inc.|           193|         186|
+-------------+--------------------+--------------+------------+
only showing top 20 rows
```

***Figure 6-18.*** *Scheduled and elapsed flight time for each flight number*

Now let's say we are analyzing this data since we are interested in the differences between the scheduled time for a flight and the actual time the flight took. While we can manually look at each of the rows in the data frame to figure out what the difference between both of the time columns is, it is far easier to let Spark perform this calculation for you. For this scenario, we are using the following code (Listing 6-21) to create a new data frame that selects a subset of the columns of the original df_flightinfo data frame and does a simple calculation between the SCHEDULE_TIME and the ELAPSED_TIME columns (Figure 6-19).

***Listing 6-21.*** Add a calculated column to a data frame

```
# Create a new df_flightinfo_times data frame from df_flightinfo
# with a new column that does a calculation between the scheduled and
elapsed time
df_flightinfo_times = df_flightinfo.withColumn("Time_diff", df_flightinfo.
ELAPSED_TIME - df_flightinfo.SCHEDULED_TIME).select("FLIGHT_NUMBER",
"AIRLINE", "SCHEDULED_TIME", "ELAPSED_TIME", "Time_diff")

# Return the first ten rows
df_flightinfo_times.show(10)
```

```
+-------------+--------------------+--------------+------------+---------+
|FLIGHT_NUMBER|             AIRLINE|SCHEDULED_TIME|ELAPSED_TIME|Time_diff|
+-------------+--------------------+--------------+------------+---------+
|           98|Alaska Airlines Inc.|           205|         194|      -11|
|         2336|American Airlines...|           280|         279|       -1|
|          840|     US Airways Inc.|           286|         293|        7|
|          258|American Airlines...|           285|         281|       -4|
|          135|Alaska Airlines Inc.|           235|         215|      -20|
|          806|Delta Air Lines Inc.|           217|         230|       13|
|          612|    Spirit Air Lines|           181|         170|      -11|
|         2013|     US Airways Inc.|           273|         249|      -24|
|         1112|American Airlines...|           195|         193|       -2|
|         1173|Delta Air Lines Inc.|           221|         203|      -18|
+-------------+--------------------+--------------+------------+---------+
only showing top 10 rows
```

***Figure 6-19.*** *df_flightinfo_times data frames that show travel time information*

As we can see from Figure 6-19, the majority of the flights in our selection of ten rows actually spend less travel time than scheduled.

While seeing this information on an individual flight is very useful, it would also be very interesting to see how all the flights in our sample performed. To get an idea on things like the average, maximum (Listing 6-22 resulting in Figure 6-20), or minimum (Listing 6-23 resulting in Figure 6-21) of time difference (Listing 6-24 resulting in Figure 6-22) between the scheduled and elapsed flight time, we can call a number of functions in PySpark.

***Listing 6-22.*** Retrieve a single aggregated value

```
# Show the maximum time diff value
df_flightinfo_times.select([max("Time_diff")]).show()
```

```
+--------------+
|max(Time_diff)|
+--------------+
|           330|
+--------------+
```

***Figure 6-20.*** *Maximum time difference between the scheduled and elapsed time*

***Listing 6-23.*** Retrieve a single aggregated value

```
# Show the minimum time diff value
df_flightinfo_times.select([min("Time_diff")]).show()
```

```
+--------------+
|min(Time_diff)|
+--------------+
|          -201|
+--------------+
```

***Figure 6-21.*** *Minimum time difference between the scheduled and elapsed time*

***Listing 6-24.*** Retrieve a single aggregated value

```
# Show the average time diff value
df_flightinfo_times.select([mean("Time_diff")]).show()
```

```
+-----------------+
|   avg(Time_diff)|
+-----------------+
|-4.887784896345963|
+-----------------+
```

***Figure 6-22.*** *Average time difference between the scheduled and elapsed time*

While it is undoubtedly helpful to know the separate commands to retrieve a number of summary statistics for a dataset, Spark also has a separate function (Listing 6-25) that can directly do that for you and combine the multiple results into a single output (Figure 6-23).

***Listing 6-25.*** Generate summary statistics from a specific column

```
# We can generate summary statistics for a specific column using a single command
df_flightinfo_times.select("Time_diff").describe().show()
```

```
+-------+------------------+
|summary|         Time_diff|
+-------+------------------+
|  count|           5714008|
|   mean|-4.887784896345963|
| stddev|12.883379307399249|
|    min|              -201|
|    max|               330|
+-------+------------------+
```

***Figure 6-23.*** *Summary statistics for the Time_diff column of the df_flightinfo_times data frame*

As we can see from the preceding summary statistics (Figure 6-23), on average the flights were performed almost 5 minutes faster than originally scheduled. We can also see there are some outliers in the data; the fastest flight arrived 201 minutes earlier than scheduled, while one of the flights took 330 minutes longer to perform than scheduled.

Perhaps we can gain some more understanding of the delays by looking at the data of flights that had a delay of more than 180 minutes. The code of Listing 6-26 selects the top 20 of those rows and sorts them based on the delay descending, meaning the flights that were delayed the most are at the top of the results (Figure 6-24).

***Listing 6-26.*** Select and sort based on a single column

```
# Select all flights that had more than 60 minutes delay
df_flightinfo_times.filter(df_flightinfo_times.Time_diff < -60).
sort(desc("Time_diff")).show(20)
```

| FLIGHT_NUMBER | AIRLINE | SCHEDULED_TIME | ELAPSED_TIME | Time_diff |
|---:|---:|---:|---:|---:|
| 1274 | Frontier Airlines... | 220 | 550 | 330 |
| 1199 | JetBlue Airways | 165 | 428 | 263 |
| 1156 | Delta Air Lines Inc. | 278 | 515 | 237 |
| 397 | United Air Lines ... | 86 | 321 | 235 |
| 1307 | American Airlines... | 253 | 484 | 231 |
| 291 | American Airlines... | 191 | 421 | 230 |
| 126 | American Airlines... | 249 | 477 | 228 |
| 718 | Spirit Air Lines | 190 | 417 | 227 |
| 1283 | American Airlines... | 226 | 446 | 220 |
| 2346 | American Airlines... | 158 | 369 | 211 |
| 4169 | Atlantic Southeas... | 78 | 286 | 208 |
| 1120 | American Airlines... | 216 | 424 | 208 |
| 6446 | Skywest Airlines ... | 130 | 336 | 206 |
| 1106 | United Air Lines ... | 99 | 304 | 205 |
| 2335 | American Airlines... | 146 | 349 | 203 |
| 125 | American Airlines... | 118 | 320 | 202 |
| 4739 | Skywest Airlines ... | 82 | 281 | 199 |
| 6017 | Atlantic Southeas... | 56 | 254 | 198 |
| 643 | American Airlines... | 75 | 272 | 197 |
| 1456 | Frontier Airlines... | 225 | 421 | 196 |

***Figure 6-24.*** *Top 20 flights with the most delay*

What we can see in these results is that the airline "American Airlines" has ran into quite some flight delays based on this dataset. But are they also the airline with the most delay on average? One way to figure that out is to calculate the average time difference for each airline and return them. We can do this by grouping the data based on the airline and calculate the average delay across all flights for each distinct airline. The code example of Listing 6-27 does just that, using the groupby function together with an aggregate option (written as agg) to supply to method on which the data needs to be grouped and on which column (Figure 6-25).

***Listing 6-27.*** Aggregate a column grouped by another column

```
# Group the Time_diff data for each airline and return the average
# difference between the scheduled and the elapsed time of a flight
df_flightinfo_times.groupby("AIRLINE").agg({"Time_diff": "mean"}).
sort(desc("avg(Time_diff)")).show()
```

```
+--------------------+-------------------+
|             AIRLINE|     avg(Time_diff)|
+--------------------+-------------------+
|Hawaiian Airlines...| 1.5531752607146145|
|Frontier Airlines...|-0.7986679986679986|
|    Spirit Air Lines|-1.4113010339169916|
|Skywest Airlines ...|-1.8904308841324933|
|Atlantic Southeas...|-2.0302189086294415|
|      US Airways Inc.| -2.374790833217487|
|Alaska Airlines Inc.| -2.695489357730738|
|American Eagle Ai...|-3.5093134283387912|
|      Virgin America| -4.255779780564263|
|     JetBlue Airways| -4.764606437136032|
|American Airlines...| -5.374734022035669|
|Southwest Airline...| -6.142219553558708|
|Delta Air Lines Inc.| -7.126546206658815|
|United Air Lines ...| -8.901461708438205|
+--------------------+-------------------+
```

***Figure 6-25.*** *Average difference between scheduled and elapsed time for each airline total over all flights*

The groupby function is very useful when you want to calculate values across the entire data frame and group them on a specific function. Besides the mean option we supplied using the agg parameter, we can also use other calculation methods like sum to calculate the totals for each grouped column, or count to count the amount of occurrences for each column value (Figure 6-26).

```
+--------------------+--------------+
|             AIRLINE|sum(Time_diff)|
+--------------------+--------------+
|Hawaiian Airlines...|        118105|
|Frontier Airlines...|        -71952|
|    Spirit Air Lines|       -162572|
|      Virgin America|       -260658|
|     US Airways Inc.|       -461239|
|Alaska Airlines Inc.|       -462112|
|American Eagle Ai...|       -978365|
|Skywest Airlines ...|      -1090427|
|Atlantic Southeas...|      -1126268|
|     JetBlue Airways|      -1248527|
|American Airlines...|      -3831836|
|United Air Lines ...|      -4519824|
|Delta Air Lines Inc.|      -6202055|
|Southwest Airline...|      -7631112|
+--------------------+--------------+
```

***Figure 6-26.*** *Total difference between scheduled and elapsed time for each airline calculated over all flights*

Another thing worth pointing out is how we passed the column that is returned by the groupby function to the sort function. Whenever a calculated column is added to the data frame, it also becomes available for selecting and sorting, and you can pass the column name into those functions.

If we continue with our flight delay investigation, we can see from the grouped average and total results that American Airlines isn't performing as badly in the delay department as we first expected. As a matter of fact, on average, their flights arrive 5 minutes earlier than planned!

We are going to return to this dataset in Chapter 7, explore it further, and even make some predictions on flight delays.

## Working with SQL Queries on Spark Data Frames

So far in this chapter, we have used functions related to data frame handling to perform actions like selecting a specific column, sorting, and grouping data. Another option we have to work with the data inside data frame is by accessing it through SQL queries directly in Spark. For those who are familiar with writing SQL code, this method might prove far easier to use than learning all the new functions (and many more we haven't touched) earlier.

Before we can write SQL queries against a data frame, we have to register it as a table structure which we can do through the code in Listing 6-28.

***Listing 6-28.*** Registering a temporary table

```
# Register the df_flightinfo data frame as a (temporary) table so we can
run SQL queries against it
df_flightinfo.registerTempTable("FlightInfoTable")
```

Now that we have registered our data frame as a (temporary) table, we can run SQL queries against it using the sqlContext command (Listing 6-29) which calls the Spark SQL module which is included in the Spark engine (Figure 6-27).

***Listing 6-29.*** Select first ten rows of a table using SQL

```
# Select the top ten rows from the FlightInfoTable for a selection of
columns
sqlContext.sql("SELECT FLIGHT_NUMBER, ORIGIN_AIRPORT, DESTINATION_AIRPORT,
ELAPSED_TIME FROM FlightInfoTable").show(10)
```

```
+-------------+--------------+-------------------+------------+
|FLIGHT_NUMBER|ORIGIN_AIRPORT|DESTINATION_AIRPORT|ELAPSED_TIME|
+-------------+--------------+-------------------+------------+
|           98|           ANC|                SEA|         194|
|         2336|           LAX|                PBI|         279|
|          840|           SFO|                CLT|         293|
|          258|           LAX|                MIA|         281|
|          135|           SEA|                ANC|         215|
|          806|           SFO|                MSP|         230|
|          612|           LAS|                MSP|         170|
|         2013|           LAX|                CLT|         249|
|         1112|           SFO|                DFW|         193|
|         1173|           LAS|                ATL|         203|
+-------------+--------------+-------------------+------------+
only showing top 10 rows
```

***Figure 6-27.*** *Top ten rows of the FlightInfoTable queried using Spark SQL*

As you can see in the preceding example, we executed a simple SELECT SQL query in which we supplied a number of columns we want to return. The Spark SQL modules process the SQL query and execute it against the table structure we created earlier. Just like the example we've shown before, we still need to supply the `.show()` function to return the results in a table-like structure.

Practically everything you can do using SQL code can be applied in Spark as well. For instance, the last example (Listing 6-30) in the previous section showed how to group data and calculate an average. We can do identical processing using a SQL query as shown in the example in Figure 6-28.

***Listing 6-30.*** Aggregate a column grouped by another column with SQL

```
# Group the flight distance for each airline and return the average flight
distance for each flight
sqlContext.sql("SELECT AIRLINE, AVG(DISTANCE) FROM FlightInfoTable GROUP BY
AIRLINE ORDER BY 'avg(Distance)' DESC").show()
```

```
+--------------------+------------------+
|             AIRLINE|     avg(DISTANCE)|
+--------------------+------------------+
|Skywest Airlines ...| 496.7721639899856|
|American Eagle Ai...|422.3154070162l003|
|      Virgin America|1405.9893220037802|
|United Air Lines ...|1271.5456844081027|
|Frontier Airlines...| 967.2148597472368|
|Southwest Airline...| 740.7113345035682|
|     JetBlue Airways|1062.1751782451095|
|     US Airways Inc.| 911.5038623153763|
|Hawaiian Airlines...| 632.5918423536816|
|Atlantic Southeas...|462.25173914335716|
|Alaska Airlines Inc.|1197.4180824363411|
|Delta Air Lines Inc.| 853.6218253392869|
|American Airlines...|1041.3392223520077|
|    Spirit Air Lines| 985.2671176275143|
+--------------------+------------------+
```

***Figure 6-28.** Average flight distance grouped for each airline*

## Reading Data from the SQL Server Master Instance

A huge advantage of SQL Server Big Data Clusters is that we have access to data stored in SQL Server instances and HDFS. So far, we have mostly worked with datasets that are stored on the HDFS filesystem, accessing them directly through Spark or creating external tables using PolyBase inside SQL Server. However, we can also access data stored inside a SQL Server database inside the Big Data Cluster directly from Spark. This can be very useful in situations where you have a part of the data stored in SQL Server and the rest on HDFS and you want to bring both together. Or perhaps you want to use the distributed processing capabilities of Spark to work with your SQL table data from a performance perspective.

Getting data stored inside the SQL Server Master Instance of your Big Data Cluster is relatively straightforward since we can connect using the SQL Server JDBC driver that is natively supported in Spark. We can use the master-0.master-svc server name to indicate we want to connect to the SQL Server Master Instance (Listing 6-31).

***Listing 6-31.*** Execute SQL Query against Master Instance

```
# Connect to the SQL Server master instance inside the Big Data Cluster
# and read data from a table into a data frame
df_sqldb_sales = spark.read.format("jdbc") \
    .option("url", "jdbc:sqlserver://master-0.master-svc;databaseName=
    AdventureWorks2014") \
    .option("dbtable", "Sales.SalesOrderDetail") \
    .option("user", "sa") \
    .option("password", "[your SA password]"). ").load()
```

The preceding code sets up a connection to our SQL Server Master Instance and connects to the AdventureWorks2014 database we created there earlier in this book. Using the "dbtable" option, we can directly map a SQL table to the data frame we are going to create using the preceding code.

After executing the code, we have a copy of the SQL table data stored inside a data frame inside our Spark cluster and we can access it like we've shown earlier (Listing 6-31 leads to Figure 6-30).

To only retrieve the first ten rows, run Listing 6-32. This will result in Figure 6-29.

***Listing 6-32.*** Retrieve first ten rows

```
df_sqldb_sales.show(10)
```

| SalesOrderID | SalesOrderDetailID | CarrierTrackingNumber | OrderQty | ProductID | SpecialOfferID | UnitPrice | UnitPriceDiscount |
|---|---|---|---|---|---|---|---|
| 43659 | 1 | 4911-403C-98 | 1 | 776 | 1 | 2024.9940 | 0.0000 |
| 43659 | 2 | 4911-403C-98 | 3 | 777 | 1 | 2024.9940 | 0.0000 |
| 43659 | 3 | 4911-403C-98 | 1 | 778 | 1 | 2024.9940 | 0.0000 |
| 43659 | 4 | 4911-403C-98 | 1 | 771 | 1 | 2039.9940 | 0.0000 |
| 43659 | 5 | 4911-403C-98 | 1 | 772 | 1 | 2039.9940 | 0.0000 |
| 43659 | 6 | 4911-403C-98 | 2 | 773 | 1 | 2039.9940 | 0.0000 |
| 43659 | 7 | 4911-403C-98 | 1 | 774 | 1 | 2039.9940 | 0.0000 |
| 43659 | 8 | 4911-403C-98 | 3 | 714 | 1 | 28.8404 | 0.0000 |
| 43659 | 9 | 4911-403C-98 | 1 | 716 | 1 | 28.8404 | 0.0000 |
| 43659 | 10 | 4911-403C-98 | 6 | 709 | 1 | 5.7000 | 0.0000 |

***Figure 6-29.*** *Data frame created from a table inside the SQL Server Master Instance*

Something that is interesting to point out for this process is the fact that Spark automatically sets the datatypes for each column to the same type as it is configured inside the SQL Server database (with some exceptions in datatype naming, datetime in SQL is timestamp in Spark, and datatypes that are not directly supported in Spark, like uniqueidentifier) which you can see in the schema of the data frame shown in Figure 6-30.

```
root
 |-- SalesOrderID: integer (nullable = true)
 |-- SalesOrderDetailID: integer (nullable = true)
 |-- CarrierTrackingNumber: string (nullable = true)
 |-- OrderQty: integer (nullable = true)
 |-- ProductID: integer (nullable = true)
 |-- SpecialOfferID: integer (nullable = true)
 |-- UnitPrice: decimal(19,4) (nullable = true)
 |-- UnitPriceDiscount: decimal(19,4) (nullable = true)
 |-- LineTotal: decimal(38,6) (nullable = true)
 |-- rowguid: string (nullable = true)
 |-- ModifiedDate: timestamp (nullable = true)
```

***Figure 6-30.*** *Data frame schema of our imported data frame from the SQL Server Master Instance*

Next to creating a data frame from a SQL table, we can also supply a query to select only the columns we are after, or perhaps perform some other SQL functions like grouping the data. The example in Listing 6-33 shows how we can load a data frame using a SQL query (Figure 6-31).

***Listing 6-33.*** Use SQL Query instead of mapping a table for a data frame

```
# While we can map a table to a data frame, we can also execute a SQL query
df_sqldb_query = spark.read.format("jdbc") \
    .option("url", "jdbc:sqlserver:// master-0.master-svc;databaseName=Adve
    ntureWorks2014") \
    .option("query", "SELECT SalesOrderID, OrderQty, UnitPrice,
    UnitPriceDiscount FROM Sales.SalesOrderDetail") \
    .option("user", "sa") \
    .option("password", "[your SA password]").load()

df_sqldb_query.printSchema()
```

```
root
 |-- SalesOrderID: integer (nullable = true)
 |-- OrderQty: integer (nullable = true)
 |-- UnitPrice: decimal(19,4) (nullable = true)
 |-- UnitPriceDiscount: decimal(19,4) (nullable = true)
```

***Figure 6-31.*** *Schema of the df_sqldb_query data frame*

# Plotting Graphs

So far, we have mostly dealt with results that are returned in a text like format whenever we execute a piece of code inside our PySpark notebook. However, when performing tasks like data exploration, it is often far more useful to look at the data in a more graphical manner. For instance, plotting histograms of your data frame can provide a wealth of information regarding the distribution of your data, while a scatter plot can help you visually understand how different columns can correlate with each other.

Thankfully, we can easily install and manage packages through Azure Data Studio that can help us plot graphs of the data that is stored inside our Spark cluster and display those graphs inside notebooks. That's not to say that plotting graphs of data that is stored inside data frames is easy. As a matter of fact, there are a number of things we need to consider before we can start plotting our data.

The first, and most important one, is that a data frame is a logical representation of our data. The actual, physical data itself is distributed across the worker nodes that make up our Spark cluster. This means that if we want to plot data through a data frame, things get complex very fast since we need to combine the data from the various nodes into a single dataset on which we can create our graph. Not only would this lead to very bad performance since we are basically removing the distributed nature of our data, but it can also potentially lead to errors since we would need to fit all of our data inside the memory of a single node. While these issues might not occur on small datasets, the larger your dataset gets, the faster you will run into these issues.

To work around these problems, we usually resort to different methods of analyzing the data. For instance, instead of analyzing the entire dataset, we can draw a sample from the dataset, which is a representation of the dataset as a whole, and plot our graphs on this smaller sample dataset. Another method can be to filter out only the data that you need, and perhaps do some calculations on it in advance, and save that as a separate, smaller, dataset before plotting it.

Whichever method you choose to create a smaller dataset for graphical exploration, one thing we will be required to do is to bring the dataset to our main Spark master node on which we submit our code. The Spark master node needs to be able to load the dataset in memory, meaning that the master node needs enough physical memory to load the dataset and not run out-of-memory and crash. One way we can do this is by converting our Spark data frame to a Pandas data frame. Pandas, which is an abbreviation for "**pan**el **da**ta," is a term that is used in the world of statistics to describe multidimensional datasets. Pandas is a Python library written for data analysis and manipulation, and if you have ever done anything with data inside Python, you are bound to have worked with it. Pandas also brings in some plotting capabilities by using the matplotlib library. While Pandas is, by default, included inside the libraries of Big Data Clusters, matplotlib isn't. The installation of the matplotlib package is however very straightforward and easy to achieve by using the "Manage Packages" option inside a notebook that is connected to your Big Data Cluster (Figure 6-32).

***Figure 6-32.** Manage Packages option inside the notebook header*

After clicking the Manage Packages button, we can see what packages are already installed and are presented an option to install additional packages through the "Add new" tab (Figure 6-33).

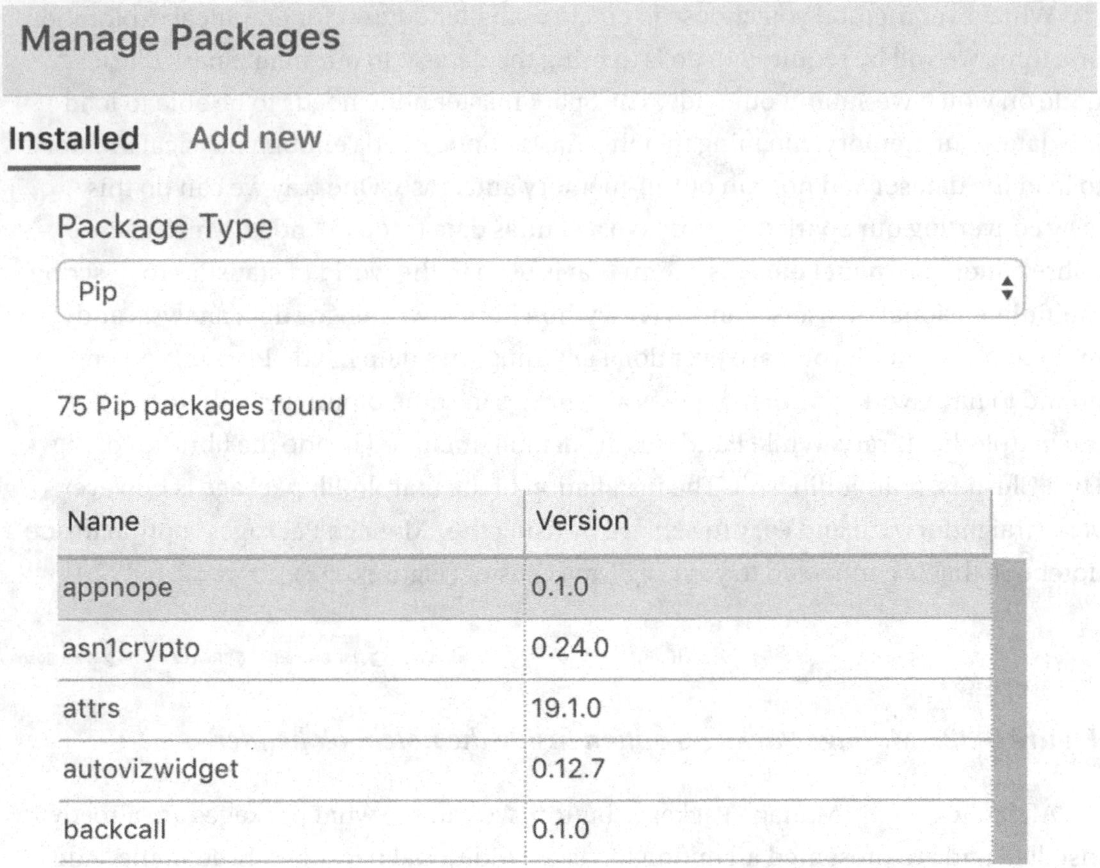

Manage Packages

Installed Add new

Package Type

Pip

75 Pip packages found

| Name | Version |
|---|---|
| appnope | 0.1.0 |
| asn1crypto | 0.24.0 |
| attrs | 19.1.0 |
| autovizwidget | 0.12.7 |
| backcall | 0.1.0 |

***Figure 6-33.*** *Manage Packages*

In this case we are going to install the matplotlib packages so we can work through the examples further on in this chapter. In Figure 6-34 I searched for the matplotlib package inside the Add new packages tab and selected the latest stable build of matplotlib currently available.

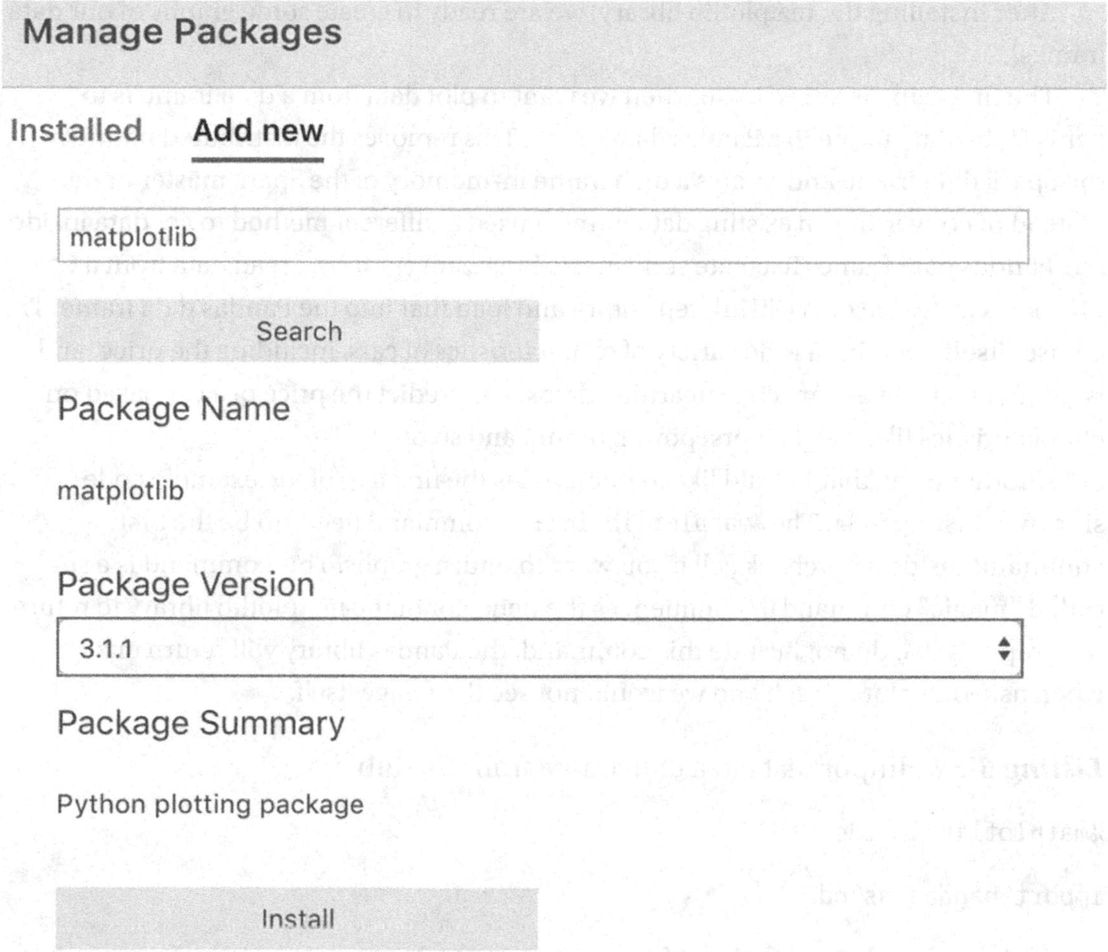

Manage Packages

Installed Add new

matplotlib

Search

Package Name

matplotlib

Package Version

3.1.1

Package Summary

Python plotting package

Install

***Figure 6-34.*** *Matplotlib package installation*

After selecting the package and the correct version, you can click the "Install" button to perform the installation of the package unto your Big Data Cluster. The installation process is visible through the "TASKS" console at the bottom area of Azure Data Studio as shown in Figure 6-35.

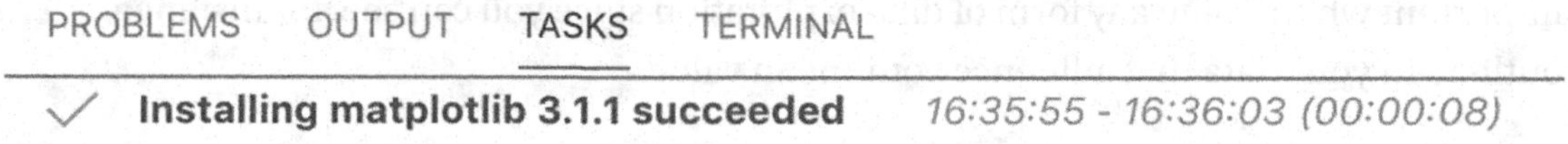

PROBLEMS OUTPUT TASKS TERMINAL

✓ **Installing matplotlib 3.1.1 succeeded** *16:35:55 - 16:36:03 (00:00:08)*

***Figure 6-35.*** *Matplotlib installation task*

After installing the matplotlib library, we are ready to create some graphs of our data frames!

The first thing we need to do when we want to plot data from a data frame is to convert the data frame to a Pandas data frame. This removes the distributed nature of the Spark data frame and creates a data frame in-memory of the Spark master node. Instead of converting an existing data frame, I used a different method to get data inside our Pandas data frame. To create some more interesting graphs, I read data from a CSV file that is available on a GitHub repository and load that into the Pandas data frame. The dataset itself contains a wide variety of characteristics of cars, including the price, and is frequently used as a machine learning dataset to predict the price of a car based on characteristics like weight, horsepower, brand, and so on.

Another thing that I would like to point out is the first line of the example code shown in Listing 6-34. The `%matplotlib inline` command needs to be the first command inside a notebook cell if you want to return graphs. This command is a so-called "magic" command that influences the behavior of the matplotlib library to return the graphs. If we do not include this command, the Pandas library will return errors when asked to plot a graph and we would not see the image itself.

***Listing 6-34.*** Import data to a data frame from GitHub

```
%matplotlib inline

import pandas as pd

# Create a local Pandas data frame from a csv through a URL
pd_data_frame = pd.read_csv("https://github.com/Evdlaar/Presentations/raw/
master/Advanced%20Analytics%20in%20the%20Cloud/automobiles.csv")
```

After running the preceding code, we can start to create graphs using the `pd_data frame` as a source.

The code of Listing 6-35 will create a histogram of the horsepower column inside our Pandas data frame (Figure 6-36) using the `hist()` function of Pandas. Histograms are incredibly useful for seeing how your data is distributed. Data distribution is very important when doing any form of data exploration since you can see, for instance, outliers in your data that influence your mean value.

***Listing 6-35.*** Create a histogram for a single column

```
%matplotlib inline

# We can create a histogram, for instance, for the horsepower column
pd_data_frame.hist("horsepower")
```

```
array([[<matplotlib.axes._subplots.AxesSubplot object at 0x123265518>]],
      dtype=object)
```

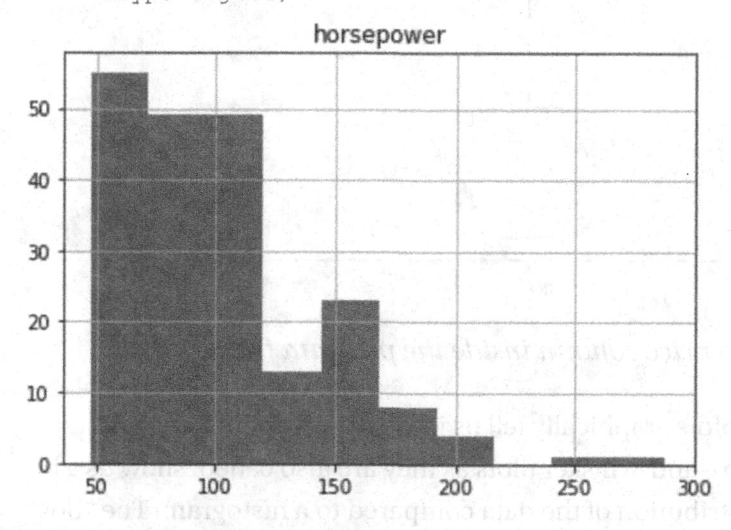


***Figure 6-36.*** *Histogram of the horsepower column of the pd_data frame Pandas data frame*

Next to histograms we can basically create any graph type we are interested in. Pandas supports many different graph types and also many options to customize how your graphs look like. A good reference for what you can do can be found on the Pandas documentation page at `https://pandas.pydata.org/pandas-docs/stable/user_guide/visualization.html`.

To give you another example of the syntax, the code of Listing 6-36 creates a boxplot of the price column inside our Pandas data frame (Figure 6-37).

***Listing 6-36.*** Generate a boxplot based on a single column

```
%matplotlib inline

# Also other graphs like boxplots are supported
# In this case we create a boxplot for the "price" column
pd_data_frame.price.plot.box()
```

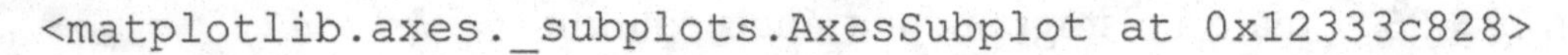

```
<matplotlib.axes._subplots.AxesSubplot at 0x12333c828>
```

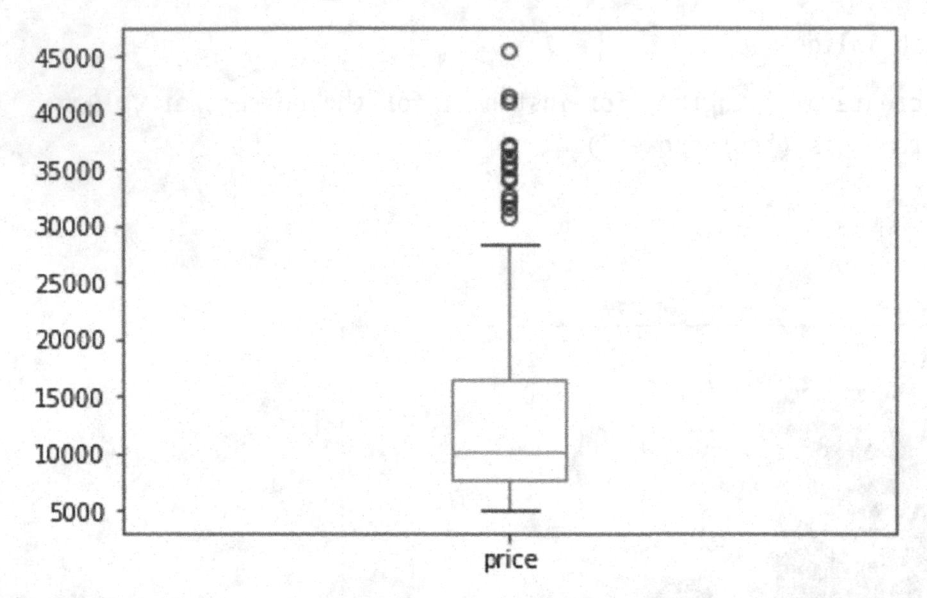


***Figure 6-37.** Boxplot of the price column inside the pd_data frame*

Just like histograms, boxplots graphically tell us information about how our data is distributed. Boxplots, or box-and-whisker plots as they are also called, show us a bit more detail regarding the distribution of the data compared to a histogram. The "box" of the boxplot is called the interquartile range (IQR) and contains the middle 50% of our data. In our case we can see that the middle 50% of our price data is somewhere between 7,500 and 17,500. The line, or whisker, beneath the IQR shows the bottom 25% of our data and the whisker above the IQR the top 25%. The circles above the top whisker show the outliers of our dataset, in the case of this example dataset, to indicate cars that are priced higher than $1.5 * $ IQR. Outliers have a potentially huge impact on the average price and are worth investigating to make sure they are not errors. Finally, the green bar inside the IQR indicates the mean, or average, price for the price column.

Boxplots are frequently used to compare the data distribution of multiple datasets against each other. Something we can also do inside our PySpark notebook by setting the `subplot()` function of the `matplotlib` library. The parameters we set for `subplot()` dictate the location, expressed in rows and columns, the plot following the `subplot()` function should be displayed in. In the example in Listing 6-37, the boxplot for the price column is shown in location `1,2,1` which means 1 row, 2 columns, first column. The plot for the horsepower is shown in location 1 row, 2 columns, second column effectively plotting both boxplots next to each other (Figure 6-38).

***Listing 6-37.*** Generate multiple boxplots to compare two values

```
%matplotlib inline

# Boxplots are frequently used to compare the distribution of datasets
# We can plot multiple boxplots together and return them as one image using
the following code
import matplotlib.pyplot as plt

plt.subplot(1, 2, 1)
pd_data frame.price.plot.box()

plt.subplot(1, 2, 2)
pd_data frame.horsepower.plot.box()

plt.show()
```

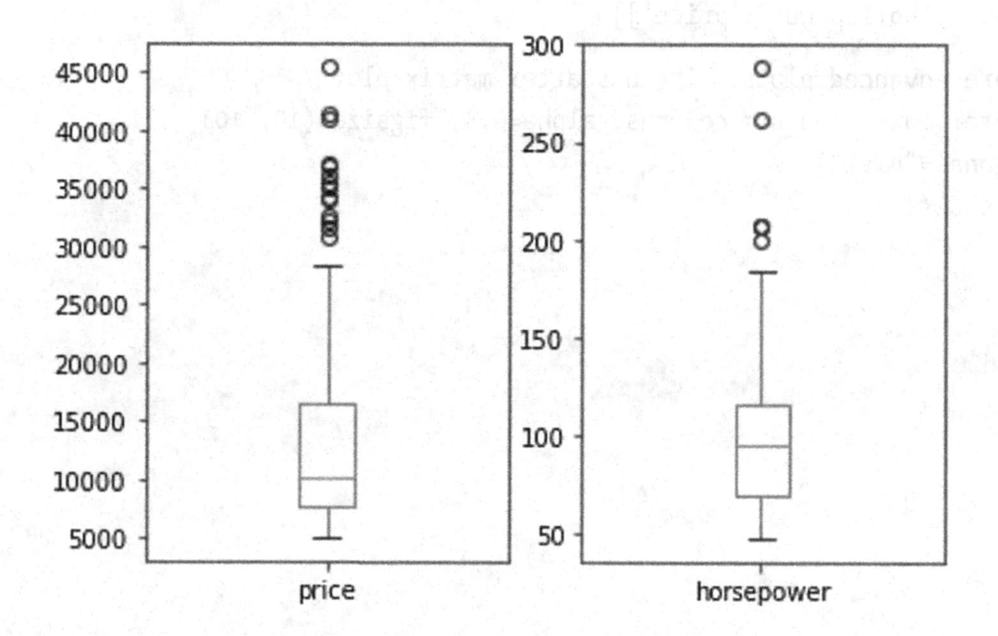


***Figure 6-38.*** *Two different boxplots plotted next to each other*

We won't go into further detail about boxplots since they are outside the scope of this book, but if you are interested in learning more about them, there are plenty of resources available online about how to interpret them.

One final example we would like to show displays how powerful the graphical capabilities of Pandas is. Using the code of Listing 6-38, we will create a so-called scatter matrix (Figure 6-39). A scatter matrix consists of many different graphs all combined into a single, large, graph. The scatter matrix returns a scatter plot for each interaction between the columns we provide and a histogram if the interaction is on an identical column.

***Listing 6-38.*** Create scatter matrix

```
%matplotlib inline

import matplotlib.pyplot as plt
from pandas.plotting import scatter_matrix

# Only select a number of numerical columns from our data frame
pd_num_columns = pd_data frame[['length','width','height','curb-
weight','horsepower','price']]

# More advanced plots, like a scatter matrix plot
scatter_matrix(pd_num_columns, alpha=0.5, figsize=(10, 10),
diagonal="hist")
plt.show()
```

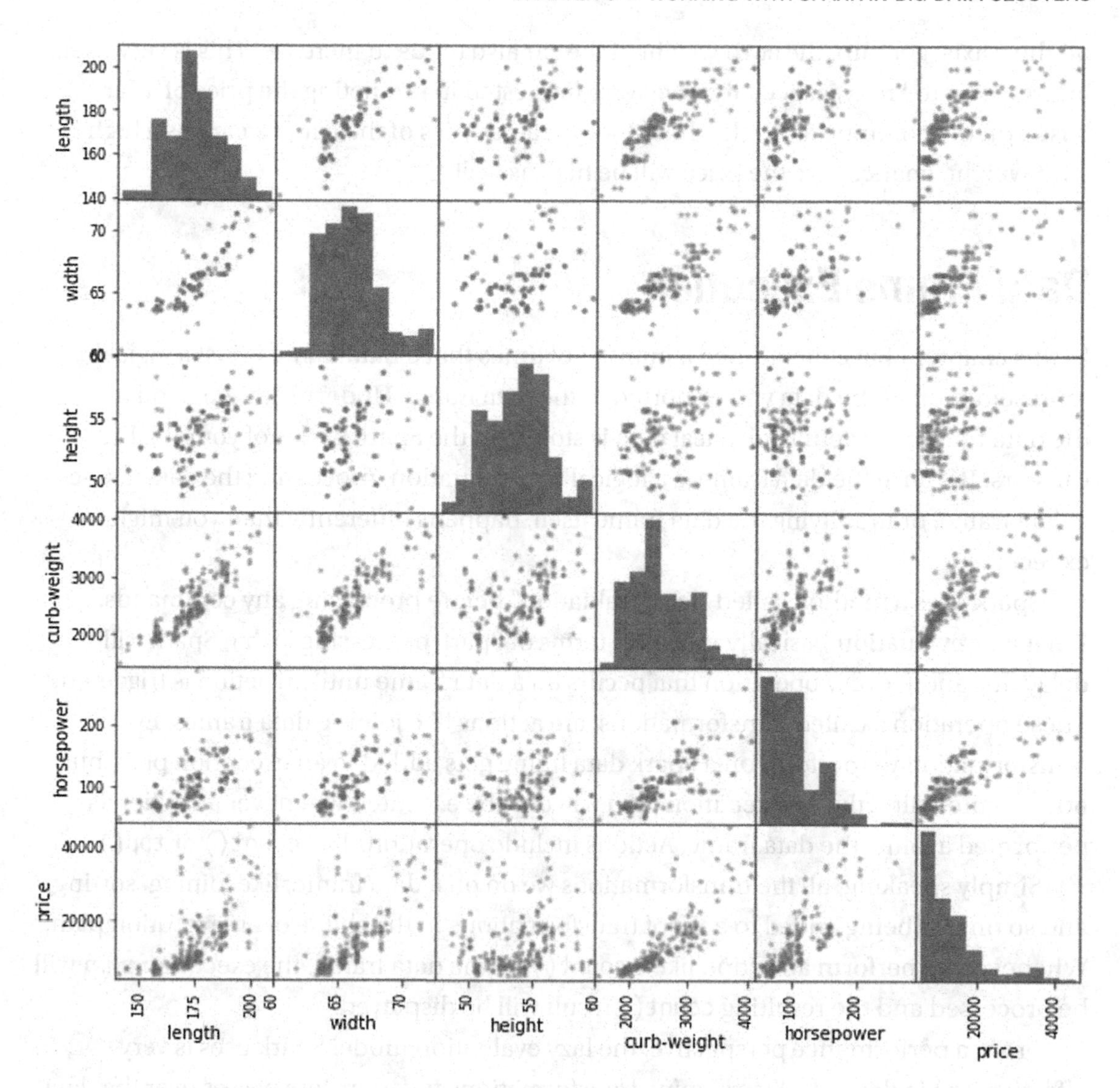


***Figure 6-39.** Scatter matrix plot on various columns inside the pd_data frame*

A scatter matrix plot is incredibly useful when you want to detect correlations between the various columns inside your dataset. Each of the scatter plots draws a dot for each value on the x axis (for instance, length) and the y axis (for instance, price). If these dots tend to group together, making up a darker dot in the case of the plot above, the data inside the columns could potentially be correlated to each other, meaning that if one has a higher or lower value, the other generally moves in the same direction. A more practical example of this is the plot between price (bottom left of the preceding graph) and curb weight (fourth from the right in the preceding graph). As the price, displayed on the y axis, goes up, the curb weight of the car also tends to increase. This is very useful information to know, especially if we were interested in predicting the price of a car based on the column values that hold the characteristics of the car. If a car has a high curb weight, chances are the price will be high as well.

## Data Frame Execution

In this chapter I have mentioned a number of times that a data frame is just a logical representation of the data you imported to the data frame. Underneath the hood of the data frame, the actual, physical data is stored on the Spark nodes of your Big Data Clusters. Because the data frame is a logical representation, processing the data inside a data frame, or modifying the data frame itself, happens differently than you might expect.

Spark uses a method called "lazy evaluation" before processing any commands. What lazy evaluation basically means in terms of Spark processing is that Spark will delay just about every operation that occurs on a data frame until an action is triggered. These operations, called transformations, are actions like joining data frames. Every transformation we perform on a Spark data frame gets added to an execution plan, but not executed directly. An execution plan will only be executed whenever an action is performed against the data frame. Actions include operations like `count()` or `top()`.

Simply speaking, all the transformations we do on a data frame, like joining, sorting, and so on, are being added to a list of transformations in the shape of an execution plan. Whenever we perform an action like a `count()` on the data frame, the execution plan will be processed and the resulting `count()` result will be displayed.

From a performance perspective, the lazy evaluation model Spark uses is very effective on big datasets. By grouping transformations together, less passes over the data are required to perform the requested operations. Also, grouping the transformations together creates room for optimization. If Spark knows all operations it needs to perform on the data, it can decide on the optimal method to perform the actions required for the end result, perhaps some operations can be avoided or others can be combined together.

From inside our PySpark notebook, we can very easily see the execution plan by using the `explain()` command. In the example in Listing 6-39, we are going to import flight and airport information into two separate data frames and look at the execution plan of one of the two data frames (Figure 6-40).

***Listing 6-39.*** Explain the execution plan of a single table data frame

```
# Import the flights and airlines data again if you haven't already
df_flights = spark.read.format('csv').options(header='true',
inferSchema='true').load('/Flight_Delays/flights.csv')
df_airlines = spark.read.format('csv').options(header='true',
inferSchema='true').load('/Flight_Delays/airlines.csv')

# Just like SQL Server, Spark uses execution plans which you can see
through .explain()
df_flights.explain()
```

```
== Physical Plan ==
*(1) FileScan csv [YEAR#10,MONTH#11,DAY#12,DAY_OF_WEEK#13,AIRLINE#14,FLIGHT_NUMBER#15,TAIL_NUMBER#16,ORIGIN_AIRPORT#1
```

***Figure 6-40.*** *Execution plan of a newly imported data frame*

As you can see in Figure 6-40, there is only a single operation so far, a FileScan, which is responsible for reading the CSV contents into the `df_flights` data frame. As a matter of fact, the data is not already loaded into the data frame when we execute the command, but it will be the first step in the execution plan whenever we perform an action to trigger the actual load of the data.

To show changes occurring to the execution plan, we are going to join both the data frames we created earlier together (Listing 6-40) and look at the plan (Figure 6-41).

***Listing 6-40.*** Explain the execution plan of a multitable data frame

```
# Let's join both data frames again and see what happens to the plan
from pyspark.sql.functions import *

df_flightinfo = df_flights.join(df_airlines, df_flights.AIRLINE == df_
airlines.IATA_CODE, how="inner").drop(df_flights.AIRLINE)

df_flightinfo.explain()
```

```
== Physical Plan ==
*(2) Project [YEAR#10, MONTH#11, DAY#12, DAY_OF_WEEK#13, FLIGHT_NUMBER#15, TAIL_NUMBER#16, ORIGIN_AIRPORT#17, DESTINA
+- *(2) BroadcastHashJoin [AIRLINE#14], [IATA_CODE#82], Inner, BuildRight
   :- *(2) Project [YEAR#10, MONTH#11, DAY#12, DAY_OF_WEEK#13, AIRLINE#14, FLIGHT_NUMBER#15, TAIL_NUMBER#16, ORIGIN_A
   :  +- *(2) Filter isnotnull(AIRLINE#14)
   :     +- *(2) FileScan csv [YEAR#10,MONTH#11,DAY#12,DAY_OF_WEEK#13,AIRLINE#14,FLIGHT_NUMBER#15,TAIL_NUMBER#16,ORIG
   +- BroadcastExchange HashedRelationBroadcastMode(List(input[0, string, true]))
      +- *(1) Project [IATA_CODE#82, AIRLINE#83]
         +- *(1) Filter isnotnull(IATA_CODE#82)
            +- *(1) FileScan csv [IATA_CODE#82,AIRLINE#83] Batched: false, Format: CSV, Location: InMemoryFileIndex[h
```

***Figure 6-41.*** *Execution plan of a data frame join*

From the execution plan, we can see two FileScan operations that will read the contents of both source CSV files into their data frames. Then we can see that Spark decided on performing a hash join on both data frames on the key columns we supplied in the preceding code.

Again, the actions we performed against the data frame have not been actually executed. We can trigger this by performing an action like a simple `count()` (Listing 6-41).

***Listing 6-41.*** Perform an action to execute the execution plan

```
# Even though we joined the data frames and see that reflected in the
# execution plan, the plan hasn't been executed yet
# Execution plans only get executed when performing actions like count() or top()
df_flightinfo.count()
```

The execution plan will still be attached to the data frame, and any subsequent transformations we perform will be added to it. Whenever we perform an action at a later point in time, the execution plan will be executed with all the transformations that are part of it.

# Data Frame Caching

One method to optimize the performance of working with data frames is to cache them. By caching a data frame, we place it inside the memory of the Spark worker nodes and thus avoid the cost of reading the data from disk whenever we perform an action against a data frame. When you need to cache, a data frame is depended on a large number of factors, but generally speaking whenever you perform multiple actions against a data frame in a single script, it is often a good idea to cache the data frame to speed up performance of subsequent actions.

We can retrieve information about whether or not (Figure 6-42) our data frame is cached by calling the `storageLevel` function as shown in the example in Listing 6-42.

***Listing 6-42.*** Retrieve the data frame's storage level

```
df_flightinfo.storageLevel
```

```
StorageLevel(False, False, False, False, 1)
```

***Figure 6-42.*** *Caching information of the df_flightinfo data frame*

The function returns a number of Boolean values on the level of caching that is active for this data frame: Disk, Memory, OffHeap, and Deserialized. By default, whenever we cache a data frame, it will be cached to both Disk and Memory.

As you can see in Figure 6-43, the `df_flightinfo` data frame is not cached at this point. We can change that by calling the `cache()` function as shown in the code in Listing 6-43.

***Listing 6-43.*** Enable caching on a data frame

```
# To cache our data frame, we just have to use the .cache() function
# The default cache level is Disk and Memory
df_flightinfo.cache()

df_flightinfo.storageLevel
```

If we look at the results of the `storageLevel` function, shown in Figure 6-43, we can see the data frame is now cached.

```
StorageLevel(True, True, False, True, 1)
```

***Figure 6-43.*** *Caching information of the df_flightinfo data frame*

Even though the storageLevel function returns that the data frame is cached, it actually isn't yet. We still need to perform an action before the actual data that makes up the data frame is retrieved and can be cached. One example of an action is a `count()`, which is shown in the code of Listing 6-44.

***Listing 6-44.*** Initialize cache by performing a count on the data frame

```
# Even though we get info back that the data frame is cached, we have to
# perform an action before it actually is cached
df_flightinfo.count()
```

Besides the `storageLevel()` command which returns limited information about the caching of a data frame, we can expose far more detail through the Yarn portal.

To get to the Yarn portal, you can use the web link to the "Spark Diagnostics and Monitoring Dashboard" which is shown in the SQL Server Big Data Cluster tab whenever you connect, or manage, a Big Data Cluster through Azure Data Studio (Figure 6-44).

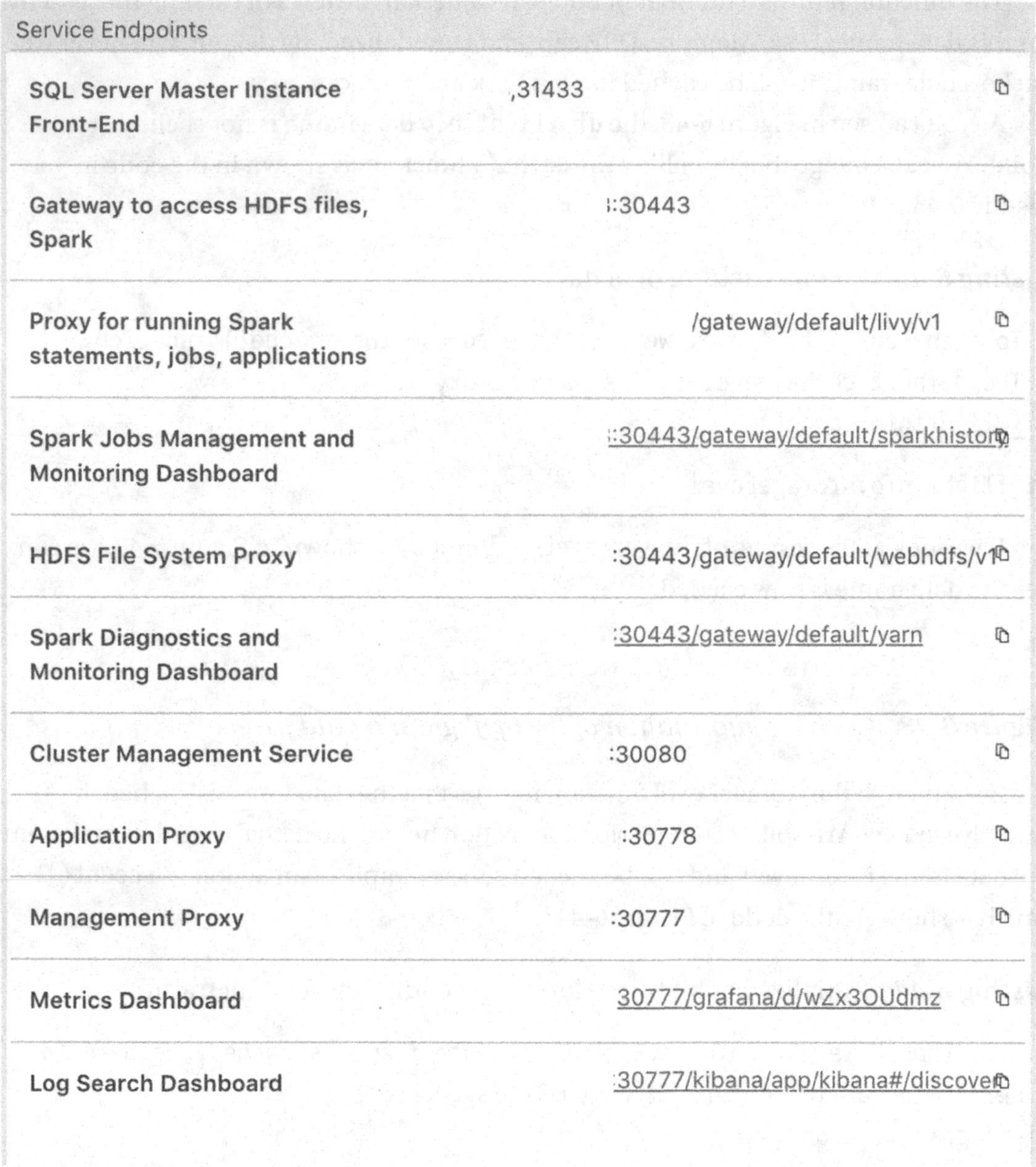

| Service Endpoints | |
|---|---|
| **SQL Server Master Instance Front-End** | ,31433 |
| **Gateway to access HDFS files, Spark** | :30443 |
| **Proxy for running Spark statements, jobs, applications** | /gateway/default/livy/v1 |
| **Spark Jobs Management and Monitoring Dashboard** | :30443/gateway/default/sparkhistory |
| **HDFS File System Proxy** | :30443/gateway/default/webhdfs/v1 |
| **Spark Diagnostics and Monitoring Dashboard** | :30443/gateway/default/yarn |
| **Cluster Management Service** | :30080 |
| **Application Proxy** | :30778 |
| **Management Proxy** | :30777 |
| **Metrics Dashboard** | 30777/grafana/d/wZx3OUdmz |
| **Log Search Dashboard** | :30777/kibana/app/kibana#/discover |

***Figure 6-44.*** *Service Endpoints in Azure Data Studio*

After logging into the Yarn web portal, we are shown an overview of all applications as shown in Figure 6-45.

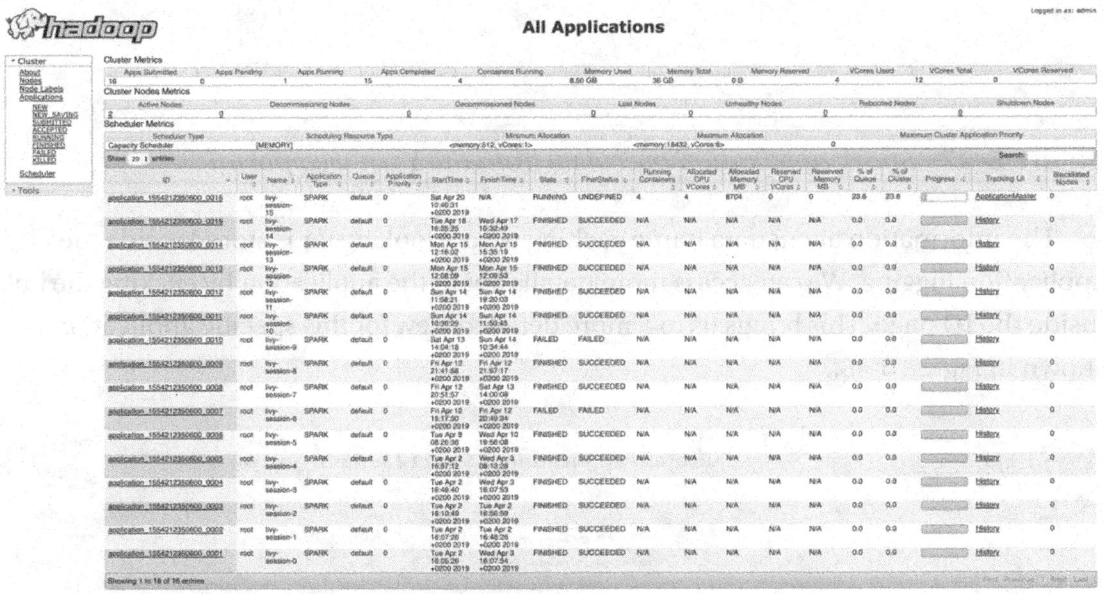

hadoop

Logged in as: admin

**All Applications**

- Cluster
  - About
  - Nodes
  - Node Labels
  - Applications
    - NEW
    - NEW_SAVING
    - SUBMITTED
    - ACCEPTED
    - RUNNING
    - FINISHED
    - FAILED
    - KILLED
  - Scheduler
- Tools

Cluster Metrics

| Apps Submitted | Apps Pending | Apps Running | Apps Completed | Containers Running | Memory Used | Memory Total | Memory Reserved | VCores Used | VCores Total | VCores Reserved |
|---|---|---|---|---|---|---|---|---|---|---|
| 16 | 0 | 1 | 15 | 4 | 8.50 GB | 36 GB | 0 B | 4 | 12 | 0 |

Cluster Nodes Metrics

| Active Nodes | Decommissioning Nodes | Decommissioned Nodes | Lost Nodes | Unhealthy Nodes | Rebooted Nodes | Shutdown Nodes |
|---|---|---|---|---|---|---|
| 2 | 0 | 0 | 0 | 0 | 0 | 0 |

Scheduler Metrics

| Scheduler Type | Scheduling Resource Type | Minimum Allocation | Maximum Allocation | Maximum Cluster Application Priority |
|---|---|---|---|---|
| Capacity Scheduler | [MEMORY] | <memory:512, vCores:1> | <memory:18432, vCores:6> | 0 |

Show 20 entries Search:

| ID | User | Name | Application Type | Queue | Application Priority | StartTime | FinishTime | State | FinalStatus | Running Containers | Allocated CPU VCores | Allocated Memory MB | Reserved CPU VCores | Reserved Memory MB | % of Queue | % of Cluster | Progress | Tracking UI | Blacklisted Nodes |
|---|---|---|---|---|---|---|---|---|---|---|---|---|---|---|---|---|---|---|---|
| application_1554212350600_0016 | root | livy-session-15 | SPARK | default | 0 | Sat Apr 20 10:40:31 +0200 2019 | N/A | RUNNING | UNDEFINED | 4 | 4 | 8704 | 0 | 0 | 23.6 | 23.6 | | ApplicationMaster | 0 |
| application_1554212350600_0015 | root | livy-session-14 | SPARK | default | 0 | Tue Apr 16 16:35:25 +0200 2019 | Wed Apr 17 10:32:49 +0200 2019 | FINISHED | SUCCEEDED | N/A | N/A | N/A | N/A | N/A | 0.0 | 0.0 | | History | 0 |
| application_1554212350600_0014 | root | livy-session-13 | SPARK | default | 0 | Mon Apr 15 12:16:02 +0200 2019 | Mon Apr 15 15:35:19 +0200 2019 | FINISHED | SUCCEEDED | N/A | N/A | N/A | N/A | N/A | 0.0 | 0.0 | | History | 0 |
| application_1554212350600_0013 | root | livy-session-12 | SPARK | default | 0 | Mon Apr 15 12:08:09 +0200 2019 | Mon Apr 15 12:09:53 +0200 2019 | FINISHED | SUCCEEDED | N/A | N/A | N/A | N/A | N/A | 0.0 | 0.0 | | History | 0 |
| application_1554212350600_0012 | root | livy-session-11 | SPARK | default | 0 | Sun Apr 14 11:58:21 +0200 2019 | Sun Apr 14 19:20:03 +0200 2019 | FINISHED | SUCCEEDED | N/A | N/A | N/A | N/A | N/A | 0.0 | 0.0 | | History | 0 |
| application_1554212350600_0011 | root | livy-session-10 | SPARK | default | 0 | Sun Apr 14 10:36:29 +0200 2019 | Sun Apr 14 11:50:45 +0200 2019 | FINISHED | SUCCEEDED | N/A | N/A | N/A | N/A | N/A | 0.0 | 0.0 | | History | 0 |
| application_1554212350600_0010 | root | livy-session-9 | SPARK | default | 0 | Sat Apr 13 14:04:18 +0200 2019 | Sun Apr 14 10:34:44 +0200 2019 | FAILED | FAILED | N/A | N/A | N/A | N/A | N/A | 0.0 | 0.0 | | History | 0 |
| application_1554212350600_0009 | root | livy-session-8 | SPARK | default | 0 | Fri Apr 12 21:41:56 +0200 2019 | Fri Apr 12 21:57:17 +0200 2019 | FINISHED | SUCCEEDED | N/A | N/A | N/A | N/A | N/A | 0.0 | 0.0 | | History | 0 |
| application_1554212350600_0008 | root | livy-session-7 | SPARK | default | 0 | Fri Apr 12 20:51:57 +0200 2019 | Sat Apr 13 14:00:08 +0200 2019 | FINISHED | SUCCEEDED | N/A | N/A | N/A | N/A | N/A | 0.0 | 0.0 | | History | 0 |
| application_1554212350600_0007 | root | livy-session-6 | SPARK | default | 0 | Fri Apr 12 15:17:50 +0200 2019 | Fri Apr 12 20:49:34 +0200 2019 | FAILED | FAILED | N/A | N/A | N/A | N/A | N/A | 0.0 | 0.0 | | History | 0 |
| application_1554212350600_0006 | root | livy-session-5 | SPARK | default | 0 | Tue Apr 9 08:26:36 +0200 2019 | Wed Apr 10 19:56:08 +0200 2019 | FINISHED | SUCCEEDED | N/A | N/A | N/A | N/A | N/A | 0.0 | 0.0 | | History | 0 |
| application_1554212350600_0005 | root | livy-session-4 | SPARK | default | 0 | Tue Apr 2 16:57:12 +0200 2019 | Wed Apr 3 08:13:28 +0200 2019 | FINISHED | SUCCEEDED | N/A | N/A | N/A | N/A | N/A | 0.0 | 0.0 | | History | 0 |
| application_1554212350600_0004 | root | livy-session-3 | SPARK | default | 0 | Tue Apr 2 16:48:40 +0200 2019 | Wed Apr 3 16:07:53 +0200 2019 | FINISHED | SUCCEEDED | N/A | N/A | N/A | N/A | N/A | 0.0 | 0.0 | | History | 0 |
| application_1554212350600_0003 | root | livy-session-2 | SPARK | default | 0 | Tue Apr 2 16:10:49 +0200 2019 | Tue Apr 2 16:56:59 +0200 2019 | FINISHED | SUCCEEDED | N/A | N/A | N/A | N/A | N/A | 0.0 | 0.0 | | History | 0 |
| application_1554212350600_0002 | root | livy-session-1 | SPARK | default | 0 | Tue Apr 2 16:07:26 +0200 2019 | Tue Apr 2 16:48:26 +0200 2019 | FINISHED | SUCCEEDED | N/A | N/A | N/A | N/A | N/A | 0.0 | 0.0 | | History | 0 |
| application_1554212350600_0001 | root | livy-session-0 | SPARK | default | 0 | Tue Apr 2 16:05:26 +0200 2019 | Wed Apr 3 16:07:54 +0200 2019 | FINISHED | SUCCEEDED | N/A | N/A | N/A | N/A | N/A | 0.0 | 0.0 | | History | 0 |

Showing 1 to 16 of 16 entries First Previous 1 Next Last

***Figure 6-45.** Yarn web portal*

Consider an application inside Spark as a unit of computation. An application can, for instance, be an interactive session with Spark through a notebook or a Spark job. Everything that we have been doing throughout this chapter inside our PySpark notebook has been processed in Spark as one or multiple applications.

As a matter of fact, the first command we execute against our Spark cluster returns information about our Spark application as you can see in Figure 6-46.

```
Starting Spark application
```

| ID | YARN Application ID | Kind | State | Spark UI | Driver log | Current session? |
|---|---|---|---|---|---|---|
| 15 | application_1554212350600_0016 | pyspark | idle | Link | Link | ✔ |

```
SparkSession available as 'spark'.
```

***Figure 6-46.** Spark application information*

For us the most important bit of information we are after is the "YARN Application ID." This ID should be present on the Yarn All Applications page, and if you are still connected to Spark through this Application ID, it should be marked as "RUNNING" like our session displayed in Figure 6-47.

| ID | User | Name | Application Type | Queue | Application Priority | StartTime | FinishTime | State | FinalStatus | Running Containers | Allocated CPU VCores | Allocated Memory MB | Reserved CPU VCores | Reserved Memory MB | % of Queue | % of Cluster |
|---|---|---|---|---|---|---|---|---|---|---|---|---|---|---|---|---|
| application_1554212350600_0016 | root | livy-session-15 | SPARK | default | 0 | Sat Apr 20 10:40:31 +0200 | N/A | RUNNING | UNDEFINED | 4 | 4 | 8704 | 0 | 0 | 23.6 | 23.6 |

***Figure 6-47.*** *Spark application overview from the Yarn web portal*

The information about data frame caching we are looking for is stored inside the application logging. We can access more details about the application by clicking the link inside the ID page. This brings us to a more detailed view for this specific application as shown in Figure 6-48.

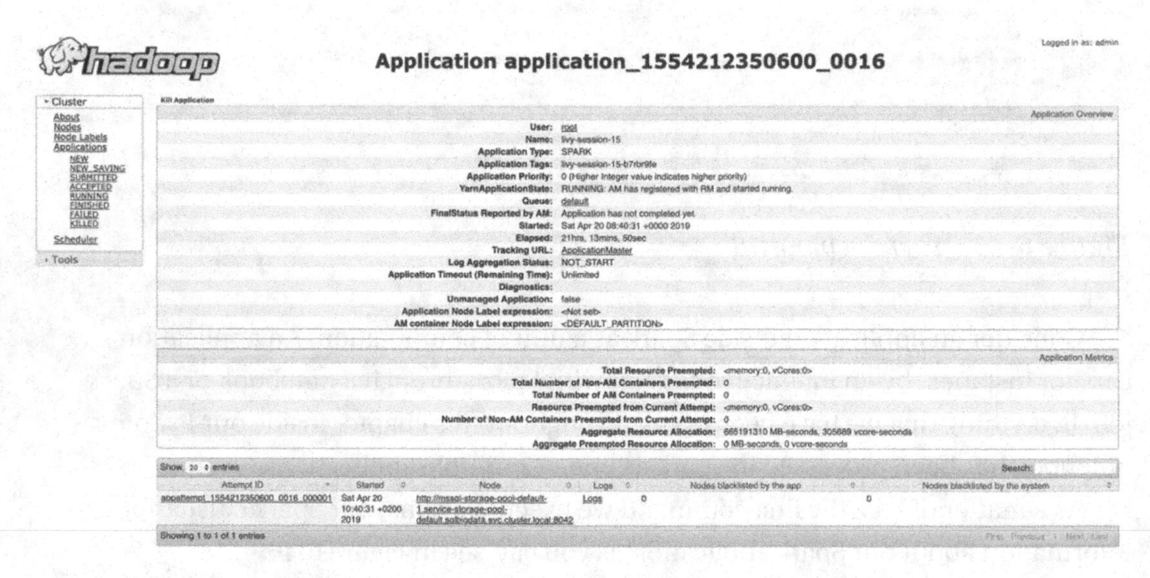


***Figure 6-48.*** *Application detail view inside the Yarn web portal*

To see the information we're after, we have to click the "ApplicationMaster" link at the "Tracking URL:" option. This opens up a web page with Spark Jobs that were, or are being, processed by this specific application. If you consider an application as your connection to the Spark cluster, a job is a command you send through your application to perform work like counting the number of rows inside a data frame. Figure 6-49 shows an overview of our Spark Jobs inside the application we are currently connected to through our PySpark notebook.

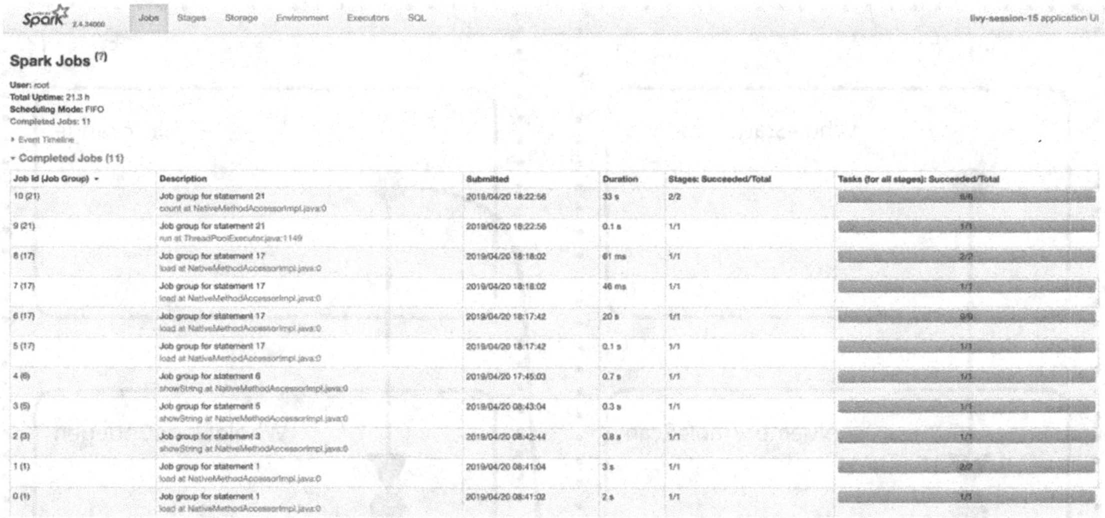

***Figure 6-49.** Spark job overview*

You can open the details of a job by clicking the link inside the "Description" column and access a wealth of information about the job processing including how the job was processed by each worker node and the Spark equivalent of the graphical execution plan for the job called the DAG (directed acyclic graph) of which an example is included in Figure 6-50.

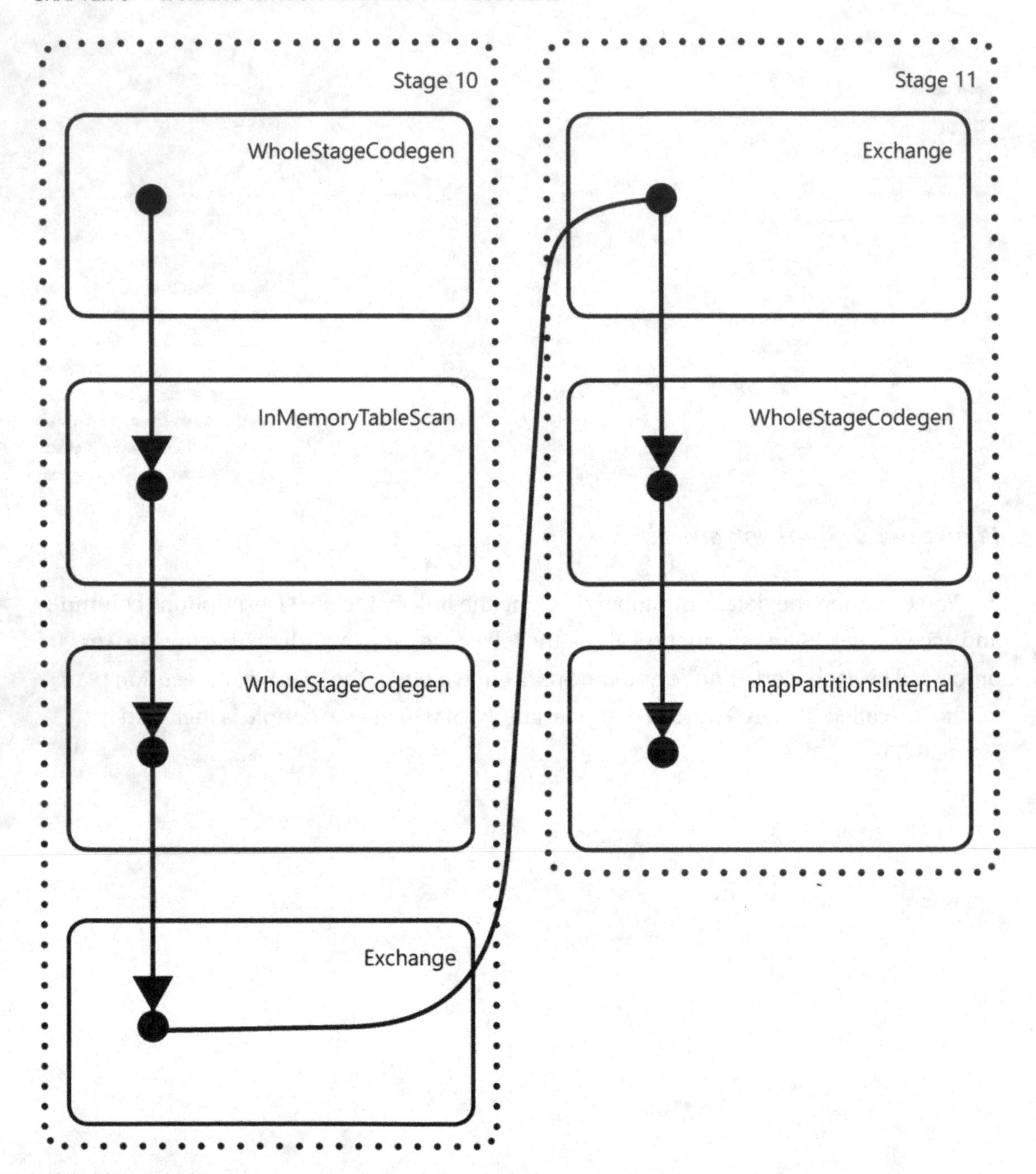


***Figure 6-50.*** *DAG of a count() function across a data frame*

To view information about data frame caching, we do not have to open the job details (though we can if we want to see storage processing for only a specific job); instead we can look at the general storage overview by clicking the "Storage" menu item at the top bar of the web page.

On this page we can see all the data frames that are currently using storage, either physical on disk, or in-memory. Figure 6-51 shows the web page on our environment after executing the `cache()` and `count()` commands we performed at the beginning of this section.

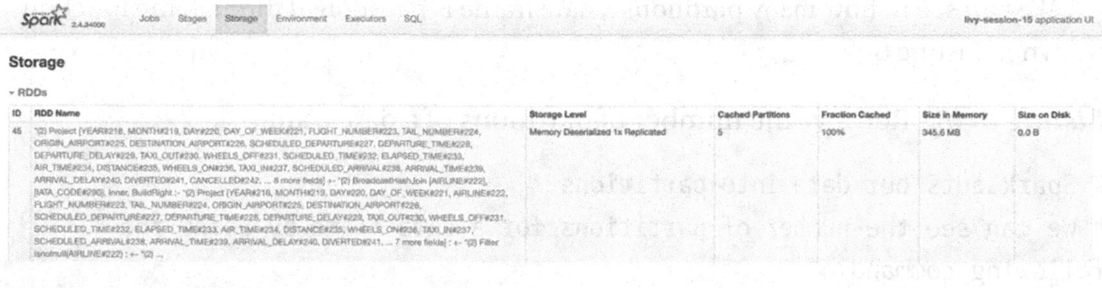

Spark 2.4.34000 Jobs Stages Storage Environment Executors SQL livy-session-15 application UI

**Storage**

▾ RDDs

| ID | RDD Name | Storage Level | Cached Partitions | Fraction Cached | Size in Memory | Size on Disk |
|---|---|---|---|---|---|---|
| 45 | *(2) Project [YEAR#218, MONTH#219, DAY#220, DAY_OF_WEEK#221, FLIGHT_NUMBER#223, TAIL_NUMBER#224, ORIGIN_AIRPORT#225, DESTINATION_AIRPORT#226, SCHEDULED_DEPARTURE#227, DEPARTURE_TIME#228, DEPARTURE_DELAY#229, TAXI_OUT#230, WHEELS_OFF#231, SCHEDULED_TIME#232, ELAPSED_TIME#233, AIR_TIME#234, DISTANCE#235, WHEELS_ON#236, TAXI_IN#237, SCHEDULED_ARRIVAL#238, ARRIVAL_TIME#239, ARRIVAL_DELAY#240, DIVERTED#241, CANCELLED#242, ... 8 more fields] +- *(2) BroadcastHashJoin [AIRLINE#222], [IATA_CODE#290], Inner, BuildRight :- *(2) Project [YEAR#218, MONTH#219, DAY#220, DAY_OF_WEEK#221, AIRLINE#222, FLIGHT_NUMBER#223, TAIL_NUMBER#224, ORIGIN_AIRPORT#225, DESTINATION_AIRPORT#226, SCHEDULED_DEPARTURE#227, DEPARTURE_TIME#228, DEPARTURE_DELAY#229, TAXI_OUT#230, WHEELS_OFF#231, SCHEDULED_TIME#232, ELAPSED_TIME#233, AIR_TIME#234, DISTANCE#235, WHEELS_ON#236, TAXI_IN#237, SCHEDULED_ARRIVAL#238, ARRIVAL_TIME#239, ARRIVAL_DELAY#240, DIVERTED#241, ... 7 more fields] : +- *(2) Filter isnotnull(AIRLINE#222) : +- *(2) ... | Memory Deserialized 1x Replicated | 5 | 100% | 345.6 MB | 0.0 B |

***Figure 6-51.** Storage usage of data frames*

What we can see from Figure 6-51 is that our data frame is completely stored in-memory, using 345.6 MB of memory spread across five partitions. We can even see on which Spark nodes the data is cached and partitioned to by clicking the link beneath the "RDD Name" column. In our case, we get back the information shown in Figure 6-52.

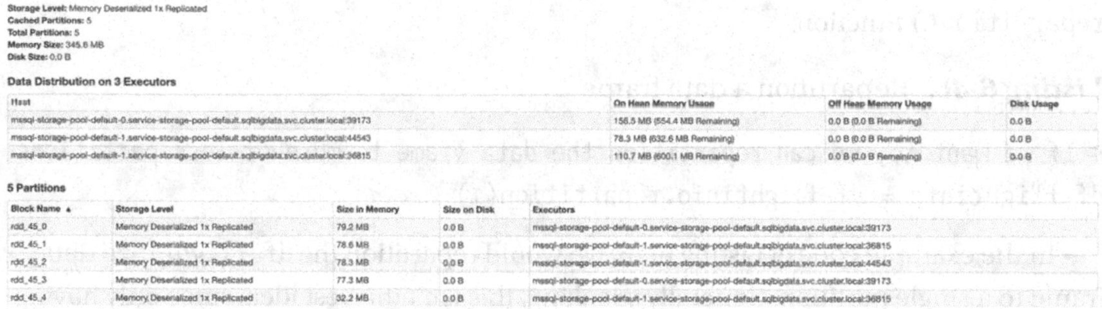

**Storage Level:** Memory Deserialized 1x Replicated
**Cached Partitions:** 5
**Total Partitions:** 5
**Memory Size:** 345.6 MB
**Disk Size:** 0.0 B

**Data Distribution on 3 Executors**

| Host | On Heap Memory Usage | Off Heap Memory Usage | Disk Usage |
|---|---|---|---|
| mssql-storage-pool-default-0.service-storage-pool-default.sqlbigdata.svc.cluster.local:39173 | 156.5 MB (554.4 MB Remaining) | 0.0 B (0.0 B Remaining) | 0.0 B |
| mssql-storage-pool-default-1.service-storage-pool-default.sqlbigdata.svc.cluster.local:44543 | 78.3 MB (632.6 MB Remaining) | 0.0 B (0.0 B Remaining) | 0.0 B |
| mssql-storage-pool-default-1.service-storage-pool-default.sqlbigdata.svc.cluster.local:36815 | 110.7 MB (600.1 MB Remaining) | 0.0 B (0.0 B Remaining) | 0.0 B |

**5 Partitions**

| Block Name ▴ | Storage Level | Size in Memory | Size on Disk | Executors |
|---|---|---|---|---|
| rdd_45_0 | Memory Deserialized 1x Replicated | 79.2 MB | 0.0 B | mssql-storage-pool-default-0.service-storage-pool-default.sqlbigdata.svc.cluster.local:39173 |
| rdd_45_1 | Memory Deserialized 1x Replicated | 78.6 MB | 0.0 B | mssql-storage-pool-default-1.service-storage-pool-default.sqlbigdata.svc.cluster.local:36815 |
| rdd_45_2 | Memory Deserialized 1x Replicated | 78.3 MB | 0.0 B | mssql-storage-pool-default-1.service-storage-pool-default.sqlbigdata.svc.cluster.local:44543 |
| rdd_45_3 | Memory Deserialized 1x Replicated | 77.3 MB | 0.0 B | mssql-storage-pool-default-0.service-storage-pool-default.sqlbigdata.svc.cluster.local:39173 |
| rdd_45_4 | Memory Deserialized 1x Replicated | 32.2 MB | 0.0 B | mssql-storage-pool-default-1.service-storage-pool-default.sqlbigdata.svc.cluster.local:36815 |

***Figure 6-52.** Storage usage of our data frame across Spark nodes*

We can see our data frame is actually cached across three Spark worker nodes, each of which cached a different amount of data. We can also see how our data frame is partitioned and how those partitions are distributed across the worker nodes. Partitioning is something that Spark handles automatically and it is essential for the distributed processing of the platform.

## Data Frame Partitioning

Like we mentioned at the end of the previous section, Spark handles the partitioning of your data frame automatically. As soon as we create a data frame, it automatically gets partitioned and those partitions are distributed across the worker nodes.

We can see in how many partitions a data frame is partitioned through the function shown in Listing 6-45.

***Listing 6-45.*** Retrieve the number of partitions of a data frame

```
# Spark cuts our data into partitions
# We can see the number of partitions for a data frame by using the
following command
df_flightinfo.rdd.getNumPartitions()
```

In our case, the `df_flightinfo` data frame we've been using throughout this chapter has been partitioned into five partitions – something we also noticed in the previous section where we looked at how the data is distributed across the Spark nodes that make up our cluster through the Yarn web portal.

If we want to, we can also set the amount of partitions ourselves (Listing 6-46). One simple way to do this is by supplying the number of partitions you want to the `repartition()` function.

***Listing 6-46.*** Repartition a data frame

```
# If we want to, we can repartition the data frame to more or less partitions
df_flightinfo = df_flightinfo.repartition(1)
```

In the example code of Listing 6-46, we would repartition the `df_flightinfo` data frame to a single partition. Generally speaking, this isn't the best idea, since only having a single partition would mean that all the processing of the data frame would end up on one single worker node. Ideally you want to partition your data frame in as equally sized partitions as possible. Whenever an action is performed against the data frame, it can get split up into equally large operations, having maximum computing efficiency.

To make sure your data frame is as efficiently partitioned as possible, it is in many cases not very efficient in just supplying the number of partitions you are interested in. In most cases you would like to partition your data on a specific key/value, making sure all rows inside your data frame that have the same key/value are partitioned together. Spark also allows partitioning on a specific column as we show in the example code of Listing 6-47.

***Listing 6-47.*** Created partitioned data frame

```
df_flights_partitioned = df_flightinfo.repartition("AIRLINE")
```

In this specific example, we are partitioning the `df_flights_partitioned` data frame on the AIRLINE column. Even though there are only 14 distinct airlines inside our data frame, we still end up with 200 partitions if we look at the partition count of the newly created data frame. That is because, by default, Spark uses a minimum partition count of 200 whenever we partition on a column. For our example, this would mean that 14 of the 200 partitions actually contain data, while the rest is empty.

Let's take a look at how the data is being partitioned and processed. Before we can do that, however, we need to perform an action against the data frame so that it actually gets partitioned (Listing 6-48).

***Listing 6-48.*** Retrieved count from partitioned data frame

```
# Let's run a count so we can get some partition information back through
the web portal
df_flights_partitioned.count()
```

After running this command, we are going to return to the Yarn web portal which we visited in the previous section when we looked at data frame caching. If you do not have it opened, you need to navigate to the Yarn web portal and open the currently running application and finally clicking the ApplicationMaster URL to view the jobs inside your application as shown in Figure 6-53.

Spark 2.4.54000 | Jobs | Stages | Storage | Environment | Executors | SQL | livy-session-15 application UI

**Spark Jobs** (?)

**User:** root
**Total Uptime:** 22.0 h
**Scheduling Mode:** FIFO
Completed Jobs: 13

▸ Event Timeline
▾ Completed Jobs (13)

| Job Id (Job Group) ▾ | Description | Submitted | Duration | Stages: Succeeded/Total | Tasks (for all stages): Succeeded/Total |
|---|---|---|---|---|---|
| 12 (24) | Job group for statement 24<br>count at NativeMethodAccessorImpl.java:0 | 2019/04/21 06:33:05 | 2 s | 3/3 | 206/206 |
| 11 (22) | Job group for statement 22<br>run at ThreadPoolExecutor.java:1149 | 2019/04/21 06:21:36 | 76 ms | 1/1 | 1/1 |
| 10 (21) | Job group for statement 21<br>count at NativeMethodAccessorImpl.java:0 | 2019/04/20 18:22:56 | 33 s | 2/2 | 6/6 |
| 9 (21) | Job group for statement 21<br>run at ThreadPoolExecutor.java:1149 | 2019/04/20 18:22:56 | 0.1 s | 1/1 | 1/1 |
| 8 (17) | Job group for statement 17<br>load at NativeMethodAccessorImpl.java:0 | 2019/04/20 18:18:02 | 61 ms | 1/1 | 2/2 |
| 7 (17) | Job group for statement 17<br>load at NativeMethodAccessorImpl.java:0 | 2019/04/20 18:18:02 | 46 ms | 1/1 | 1/1 |
| 6 (17) | Job group for statement 17<br>load at NativeMethodAccessorImpl.java:0 | 2019/04/20 18:17:42 | 20 s | 1/1 | 9/9 |
| 5 (17) | Job group for statement 17<br>load at NativeMethodAccessorImpl.java:0 | 2019/04/20 18:17:42 | 0.1 s | 1/1 | 1/1 |
| 4 (6) | Job group for statement 6<br>showString at NativeMethodAccessorImpl.java:0 | 2019/04/20 17:45:03 | 0.7 s | 1/1 | 1/1 |
| 3 (5) | Job group for statement 5<br>showString at NativeMethodAccessorImpl.java:0 | 2019/04/20 08:43:04 | 0.3 s | 1/1 | 1/1 |
| 2 (3) | Job group for statement 3<br>showString at NativeMethodAccessorImpl.java:0 | 2019/04/20 08:42:44 | 0.8 s | 1/1 | 1/1 |
| 1 (1) | Job group for statement 1<br>load at NativeMethodAccessorImpl.java:0 | 2019/04/20 08:41:04 | 3 s | 1/1 | 2/2 |
| 0 (1) | Job group for statement 1<br>load at NativeMethodAccessorImpl.java:0 | 2019/04/20 08:41:02 | 2 s | 1/1 | 1/1 |

***Figure 6-53.** Spark jobs inside our active application*

We are going to focus at the topmost job (Figure 6-54) of the list shown in Figure 6-53. This is the count we performed after manually partitioning our data frame on the AIRLINE column which you can also see in the name of the operation that was performed.

| Job Id (Job Group) ▾ | Description |
|---|---|
| 12 (24) | Job group for statement 24<br>count at NativeMethodAccessorImpl.java:0 |

***Figure 6-54.** Spark job for our count() operation*

By clicking the link in the Description, we are brought to a page that shows more information about that specific job which is shown in Figure 6-55.

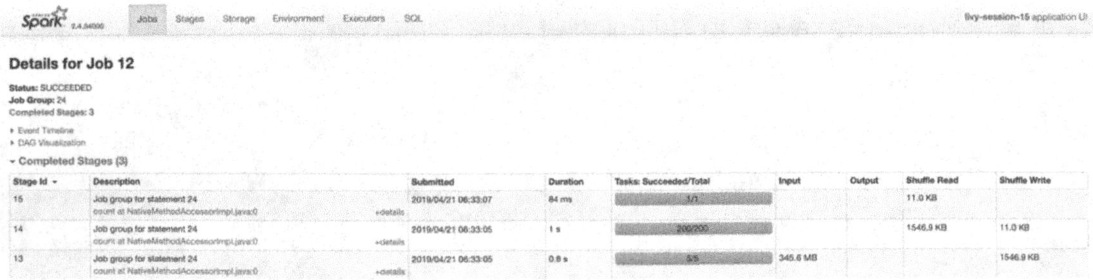

Spark 2.4.54000 | Jobs | Stages | Storage | Environment | Executors | SQL | livy-session-15 application UI

**Details for Job 12**

**Status:** SUCCEEDED
**Job Group:** 24
Completed Stages: 3

▸ Event Timeline
▸ DAG Visualization
▾ Completed Stages (3)

| Stage Id ▾ | Description | Submitted | Duration | Tasks: Succeeded/Total | Input | Output | Shuffle Read | Shuffle Write |
|---|---|---|---|---|---|---|---|---|
| 15 | Job group for statement 24<br>count at NativeMethodAccessorImpl.java:0 +details | 2019/04/21 06:33:07 | 84 ms | 1/1 | | | 11.0 KB | |
| 14 | Job group for statement 24<br>count at NativeMethodAccessorImpl.java:0 +details | 2019/04/21 06:33:05 | 1 s | 200/200 | | | 1546.9 KB | 11.0 KB |
| 13 | Job group for statement 24<br>count at NativeMethodAccessorImpl.java:0 +details | 2019/04/21 06:33:05 | 0.8 s | 5/5 | 345.6 MB | | | 1546.9 KB |

***Figure 6-55.** Spark Job details*

What is very interesting to see is that the jobs themselves are also divided into substeps called "Stages." Each stage inside a job performs a specific function. To show how our partitioning was handled, the most interesting stage is the middle one on which we zoom in in Figure 6-56.

***Figure 6-56.*** *Stage inside a Spark job*

In this stage, the actual count was performed across all the partitions of the data frame; remember, there were 200 partitions that were created when we created our partition key on the AIRLINE column. In the "Tasks: Succeeded/Total" column, you see that number being returned.

If we go down even deeper in the details of this stage, by clicking the link inside the Description column, we receive another page that shows us exactly how the data was processed for this specific stage. While this page provides a wealth of information, including an event timeline, another DAG visualization, and summary metrics for all the 200 steps (which are again called tasks on this level), I mostly want to focus on the table at the bottom of the page that returns processing information about the 200 tasks beneath this stage.

If we sort the table on the column "Shuffle Read / Records" in a descending manner, we can exactly see how many records were read from each partition for that task and from which host they were read as shown in Figure 6-57, which shows the first couple of tasks that processed rows of the total of 14 tasks that actually handled rows (the other 186 partitions are empty; thus no rows are processed from them).

- Tasks (200)

Page: 1 2 > 2 Pages. Jump to 1 . Show 100 items in a page. Go

| Index | ID | Attempt | Status | Locality Level | Executor ID | Host | Launch Time | Duration | GC Time | Shuffle Read Size / Records ▾ | Write Time | Shuffle Write Size / Records | Errors |
|---|---|---|---|---|---|---|---|---|---|---|---|---|---|
| 115 | 37 | 0 | SUCCESS | NODE_LOCAL | 1 | mssql-storage-pool-default-1.service-storage-pool-default.sqlbigdata.svc.cluster.local stdout stderr | 2019/04/21 06:33:06 | 0.2 s | | 333.2 KB / 1261855 | | 59.0 B / 1 | |
| 161 | 43 | 0 | SUCCESS | NODE_LOCAL | 1 | mssql-storage-pool-default-1.service-storage-pool-default.sqlbigdata.svc.cluster.local stdout stderr | 2019/04/21 06:33:06 | 0.1 s | | 228.6 KB / 875861 | | 59.0 B / 1 | |
| 171 | 44 | 0 | SUCCESS | NODE_LOCAL | 3 | mssql-storage-pool-default-1.service-storage-pool-default.sqlbigdata.svc.cluster.local stdout stderr | 2019/04/21 06:33:06 | 0.1 s | | 191.9 KB / 725984 | | 59.0 B / 1 | |
| 154 | 41 | 0 | SUCCESS | NODE_LOCAL | 2 | mssql-storage-pool-default-0.service-storage-pool-default.sqlbigdata.svc.cluster.local stdout stderr | 2019/04/21 06:33:06 | 94 ms | | 182.6 KB / 571977 | | 59.0 B / 1 | |

***Figure 6-57.*** *Tasks that occurred beneath our count step, sorted on Shuffle Read Size / Records*

From the results in Figure 6-57, we can immediately also see a drawback of setting our partitioning on a column value. The biggest partition contains far more rows (1,261,855) than the smallest one (61,903), meaning most of the actions we perform will occur on the Spark worker that contains our largest partition. Ideally, you want to make your partitions as even as possible and distributed in such a way that work is spread evenly across your Spark worker nodes.

# Summary

In this chapter, we took a detailed look at working with data inside the Spark architecture that is available in SQL Server Big Data Clusters.

Besides exploring the programming language PySpark to work with data frames inside Spark, we also looked at more advanced methods like plotting data. Finally, we looked a bit underneath the hood of Spark data frame processing by looking at execution plans, caching, and partitioning while introducing the Yarn web portal which provides a wealth of information about how Spark processes our data frames.

With all that data now on hand within our Big Data Cluster, let's move on to Chapter 7 to take a look at machine learning in the Big Data Cluster environment!

# CHAPTER 7

# Machine Learning on Big Data Clusters

In the previous chapters, we spent significant time on how we can query data stored inside SQL Server instances or on HDFS through Spark. One advantage of having access to data stored in different formats is that it allows you to perform analysis of the data at a large, and distributed, scale. One of the more powerful options we can utilize inside Big Data Clusters is the ability to implement machine learning solutions on our data. Because Big Data Clusters allow us to store massive amounts of data in all kinds of formats and sizes, the ability to train, and utilize, machine learning models across all of that data becomes far easier.

In many situations where you are working with machine learning, the challenge to get all the data you need to build your models on in one place takes up the bulk of the work. Building a machine learning model (or training as it is called in the data science world) becomes far easier if you can directly access all the data you require without having to move it from different data sources to one place. Besides having access to the data from a single point of entry, Big Data Clusters also allow you to operationalize your machine learning models at the same location where your data resides. This means that, technically, you can use your machine learning models to score new data as it is stored inside your Big Data Cluster. This greatly increases the capabilities of implementing machine learning inside your organization since Big Data Clusters allow you to train, exploit, and store machine learning models inside a single solution instead of having various platforms in place to perform a specific action inside your organization's advanced analytics platform.

In this chapter we are going to take a closer look at the various options available inside Big Data Clusters to train, store, and operationalize machine learning models. Generally speaking, there are two directions we are going to cover: In-Database Machine Learning Services inside SQL Server and machine learning on top of the Spark platform.




B. Weissman and E. van de Laar, *SQL Server Big Data Clusters*,
https://doi.org/10.1007/978-1-4842-5985-6_7

Both of these areas cover different use cases, but they can also overlap. As you have seen in the previous chapter, we can easily bring data stored inside a SQL Server instance to Spark and vice versa if we so please. The choice of which area you choose to perform your machine learning processes on is, in this situation, more based on what solution you personally prefer to work in. We will discuss the various technical advantages and disadvantages of both machine learning surfaces inside Big Data Clusters in each section of this chapter. This will give you a better understanding of how each of these solutions works and hopefully will help you select which one fits your requirements the best.

## SQL Server In-Database Machine Learning Services

With the release of SQL Server 2016, Microsoft introduced a new feature named in-database R Services. This new feature allows you to execute R programming code directly inside SQL Server queries using either the new `sp_execute_external_script` stored procedure or the `sp_rxPredict` CLR procedure. The introduction of in-database R Services was a new direction that allowed organizations to integrate their machine learning models directly inside their SQL Server databases by allowing the user to train, score, and store models directly inside SQL Server. While R was the only language available inside SQL Server 2016 for use with `sp_execute_external_script`, Python was added with the release of SQL Server 2017 which also resulted in a rename of the feature to Machine Learning Services. With the release of SQL Server 2019, on which Big Data Clusters are built, Java was also added as the third programming language that is available to access directly from T-SQL code.

While there are some restrictions in place regarding In-Database Machine Learning Services (for instance, some functions that are available with In-Database Machine Learning Services, like `PREDICT`, only accept algorithms developed by Revolution Analytics machine learning models), it is a very useful feature if you want to train and score your machine learning models very closely to where your data is stored. This is also the area where we believe In-Database Machine Learning Services shine. By utilizing the feature data movement is practically minimal (considering that the data your machine learning models require is also directly available in the SQL Server instance), model management is taken care of by storing the models inside SQL Server tables, and it opens the door for (near) real-time model scoring by passing the data to your machine learning model before it is stored inside a table in your database.

All of the example code inside this chapter is available as a T-SQL notebook at this book's GitHub page. For the examples in this section, we have chosen to use R as the language of choice instead of Python which we used in the previous chapter.

# Training Machine Learning Models in the SQL Server Master Instance

Before we can get started training our machine learning models, we have to enable the option to allow the use of the `sp_execute_external_script` function inside the SQL Server Master Instance of the Big Data Cluster. If you do not enable the option to run external scripts inside the SQL Instance, a large portion of the functionality of In-Database Machine Learning Services is disabled.

---

Some In-Database Machine Learning functionality is still with external scripts disabled. For instance, you can still use the PREDICT function together with a pretrained machine learning model to score data. However, you cannot run the code needed to train the model, since that mostly happens through the external script functionality.

---

If you do not have external scripts enabled and want to run a section of R code using `sp_execute_external_script`, you will be confronted with the following error message (Figure 7-1).

```
Msg 39023, Level 16, State 1, Procedure sp_execute_external_script, Line 1
'sp_execute_external_script' is disabled on this instance of SQL Server. Use sp_configure 'external scripts
enabled' to enable it.
```

***Figure 7-1.*** *Error running sp_execute_external_script with external scripts disabled*

Enabling `sp_execute_external_script` is simple and straightforward. Connect to your SQL Server Master Instance and run the code shown in Listing 7-1 to immediately enable the option.

***Listing 7-1.*** Enable external scripts

```
-- Before we can start, we need to enable external scripts
EXEC sp_configure 'external scripts enabled',1
RECONFIGURE WITH OVERRIDE
GO
```

After enabling the use of external scripts, we can directly run R, Python, or Java code through the `sp_execute_external_script` procedure. Like we mentioned in the introduction of this section, we have chosen to use R as the language of choice for this section of the book, and the code of Listing 7-2 shows a simple R command to return the version information of R.

***Listing 7-2.*** Sample R code using sp_execute_external_script

```
EXEC sp_execute_external_script
    @language =N'R',
    @script=N'print (version)'
```

Running the code in Listing 7-2 should return the results shown in Figure 7-2.

```
Started executing query at Line 1
STDOUT message(s) from external script:
NULL
               _
platform       x86_64-pc-linux-gnu
arch           x86_64
os             linux-gnu
system         x86_64, linux-gnu
status
major          3
minor          5.2
year           2018
month          12
day            20
svn rev        75870
language       R
version.string R version 3.5.2 (2018-12-20)
nickname       Eggshell Igloo


Total execution time: 00:00:04.479
```

***Figure 7-2.*** *R version results through sp_execute_external_script*

As you can see from the code, the `sp_execute_external_script` procedure accepts a number of parameters. Our example displays the minimal parameters that need to be supplied when calling the procedure, namely, `@language` and `@script`. The `@language` parameter sets the language that is used in the `@script` section. In our case, this is R. Through the `@script` parameter, we can run the R code we want to execute, in this case the `print (version)` command.

While `sp_execute_external_script` always returns results regarding the machine it is executed on, the output of the `print (version)` R command starts on line 5 with `_ platform x86_64`.

While we can work just fine with R output being returned inside the message window, we can also supply additional parameters to `sp_execute_external_script` to return the output generated with R to a table format. We do that by mapping a variable we defined in R (using the `@output_data_1_name` parameter shown in the following) to a variable we define in T-SQL and using the `WITH RESULT SETS` statement when we call the procedure as shown in the example of Listing 7-3.

***Listing 7-3.*** Returning data using WITH RESULT SETS

```
EXEC sp_execute_external_script
    @language =N'R',
    @script=N'
            r_hi <- "Hello World!"
            r_hello <- as.data.frame(r_hi)',
    @output_data_1_name = N'r_hello'
WITH RESULT SETS (([hello] nvarchar(250)));
GO
```

By running the code in Listing 7-3, you should get the text “Hello World!” returned inside a table result as shown in Figure 7-3.

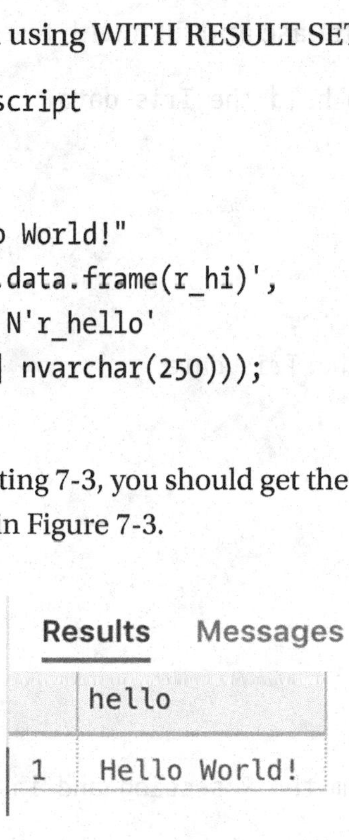

Results Messages

| | hello |
|---|---|
| 1 | Hello World! |

***Figure 7-3.*** *Output returned to a table format*

Just like how we can define and map output results through the sp_execute_external_script procedure, we can define input datasets. This is of course incredibly useful since this allows us to define a query as an input dataset to the R session and map it to an R variable. Being able to get data stored inside your SQL Server database inside the In-Database Machine Learning Service feature opens up the door to perform advanced analytics on that data like training or score machine learning models.

We are going to train a machine learning model on the "Iris" dataset. This dataset is directly available inside R and shows various characteristics of Iris flowers and to which species a specific Iris flower belongs. We can use this data to create a classification machine learning model in which we are going to predict which species an Iris flower belongs to.

Since the dataset is already present inside R, we can use a bit of R scripting together with the sp_execute_external_script procedure to return the dataset as a SQL table. The code of Listing 7-4 creates a new database called "InDBML" and a new table called "Iris" and fills that table from the Iris dataset inside an R session.

***Listing 7-4.*** Create a new database and fill it with test data

```
-- Create a new database to hold the Iris data
CREATE DATABASE InDBML
GO

USE [InDBML]
GO

-- Create a table to hold the Iris data
CREATE TABLE Iris
    (
        Sepal_Length FLOAT,
        Sepal_Width FLOAT,
        Petal_Length FLOAT,
        Petal_Width FLOAT,
        Species VARCHAR(50)
    )

-- Get the Iris dataset from the R session and insert it into our table
INSERT INTO Iris
```

```
EXEC sp_execute_external_script
    @language =N'R',
    @script=N'
            r_iris <- iris',
    @output_data_1_name = N'r_iris'

-- Get data from the new table
SELECT * FROM Iris
```

If everything processed correctly, you should have received the results as shown in Figure 7-4 which shows the values stored inside the Iris table.

Results Messages

| | Sepal_Length | Sepal_Width | Petal_Length | Petal_Width | Species |
|---|---|---|---|---|---|
| 1 | 5.1 | 3.5 | 1.4 | 0.2 | setosa |
| 2 | 4.9 | 3 | 1.4 | 0.2 | setosa |
| 3 | 4.7 | 3.2 | 1.3 | 0.2 | setosa |
| 4 | 4.6 | 3.1 | 1.5 | 0.2 | setosa |
| 5 | 5 | 3.6 | 1.4 | 0.2 | setosa |
| 6 | 5.4 | 3.9 | 1.7 | 0.4 | setosa |
| 7 | 4.6 | 3.4 | 1.4 | 0.3 | setosa |
| 8 | 5 | 3.4 | 1.5 | 0.2 | setosa |
| 9 | 4.4 | 2.9 | 1.4 | 0.2 | setosa |
| 10 | 4.9 | 3.1 | 1.5 | 0.1 | setosa |

***Figure 7-4.*** *Iris table values*

Now that we have some data to create a machine learning model on, we can get started by training a model. But before we do that, we are going to perform two additional tasks. We are going to create a "Model" table. One very useful feature of In-Database Machine Learning Services is the ability to "serialize" a model into a binary string which we can then store inside a SQL table. When the model is stored inside a table, we can retrieve it whenever we need it through a SQL query. The code of Listing 7-5 creates a model table inside the InDBML database.

***Listing 7-5.*** Create model table

```
-- Create a table to hold our trained ML models
CREATE TABLE models
    (
    model_name nvarchar(100) not null,
    model_version nvarchar(100) not null,
    model_object varbinary(max) not null
    )
 GO
```

Next to the `model_object` column that is going to hold our serialized model, we also create two additional columns that store the name and the version of the model. This can be very useful in situation where you are storing multiple models inside your SQL Server database and want to select a specific model version or name.

The next thing we are going to do is to split our Iris dataset into a training and a testing set. Splitting a dataset is a common task when you are training machine learning models. The training dataset is the data you are going to use to feed into the model you are training; the test dataset is a portion of the data you are "hiding" from the model while it is training. In that way the model was never exposed to the testing data, which means we can use the data inside the testing set to validate how well the model performs when shown data is has never seen before. For that reason, it is very important that both the training and the testing datasets are a good representation of the full dataset. For instance, if we train the model only on characteristics of the "Setosa" Iris species inside our dataset and then show it data from another species through our test dataset, it will predict wrong (predicting Setosa) since it has never seen that other species during training.

The code of Listing 7-6 randomly selects 80% of the rows from the Iris table and inserts them into a new `Iris_train` table. The other 20% of the data goes into a new `Iris_test` table.

***Listing 7-6.*** Split dataset into training and testing dataset

```
-- Randomly select 80% of the data into a separate training table
SELECT TOP 80 PERCENT *
INTO Iris_train
FROM Iris
ORDER BY NEWID()

-- Select the remaining rows into a testing table
SELECT *
INTO Iris_test
FROM Iris
EXCEPT
SELECT * FROM Iris_train
```

Now that we have a model table and got our training and testing data separated, we are ready to train our machine learning model and store it inside our model table after training which is exactly what the code of Listing 7-7 does.

***Listing 7-7.*** Train a machine learning model using sp_execute_external_script

```
DECLARE @model VARBINARY(MAX)

-- Train a decision tree based on our training dataset
EXEC sp_execute_external_script
    @language = N'R',
    @script = N'
            iris.dtree <- rxDTree(Species ~ Sepal_Length + Sepal_Width +
            Petal_Length + Petal_Width, data = iris_sqldata)

            trained_model <- rxSerializeModel(iris.dtree,
            realtimeScoringOnly = FALSE)',
    @input_data_1 = N'SELECT * FROM Iris_train',
    @input_data_1_name = N'iris_sqldata',
    @params = N'@trained_model VARBINARY(MAX) OUTPUT',
    @trained_model = @model OUTPUT

-- Insert the model into our model table
INSERT INTO models
```

```
    (
    model_name,
    model_version,
    model_object
    )
VALUES
    (
    'iris.dtree',
    'v1.0',
    @model
    )
```

The preceding code performs a number of steps to train and store a machine learning model. To make sure you understand how `sp_execute_external_script` can be used to train and store models inside your SQL Server Master Instance, we are going to describe each step that is being performed in the preceding code.

1. The first line of the script, `DECLARE @model VARBINARY(MAX)`, declares a T-SQL variable of the VARBINARY datatype that will hold our model after training it.

2. In the second step, we execute the `sp_execute_external_script` procedure and supply the R code needed to train our model. Notice we are using an algorithm called rxDTree. rxDTree is a decision tree algorithm building by Revolution Analytics, a company that Microsoft bought in 2015 and provided parallel and chunk-based algorithms for machine learning. The syntax for the model training is pretty straightforward; we are predicting the species based on the other columns (or as they are called: features) of the training dataset.

   The line `trained_model <- rxSerializeModel(iris.dtree, realtimeScoringOnly = FALSE)` is the command to serialize our model and store inside the `trained_model` R variable. We map that variable as an output parameter to the T-SQL `@model` variable in the call to the `sp_execute_external_script` procedure. We map the query that selects all the records from the training dataset as an input variable for R to use as input for the algorithm.

3. Finally, in the last step, we insert the trained model inside the model table we created earlier. We supply some additional data like a name and a version so we can easily select this model when we use it to predict Iris species in the next step.

After running this code, which should only take a few seconds, we should end up with our model stored as a binary string inside our model table as shown in Figure 7-5.

| | model_name | model_version | model_object |
|---|---|---|---|
| 1 | iris.dtree | v1.0 | 0x626C6F62C56C88687C5710DAABFE57C559E3EBFD2D1624514151A564642F… |

***Figure 7-5.*** *Trained decision tree model inside the model table*

## Scoring Data Using In-Database Machine Learning Models

Now that we have trained our model, we can use it to score, or predict, the data we stored in the Iris_test table. To do that we can use two methods, one using the `sp_execute_external_script` procedure which we have also used to train our model and the other by using the `PREDICT` function that is available in SQL Server.

The code of Listing 7-8 shows the first approach; notice that the syntax is mostly the same as the earlier examples of this method, but this time we supply the trained model as an input parameter together with a query to select the data from the Iris_test table.

***Listing 7-8.*** Run a prediction using the in-database stored model

```
-- Retrieve the model from the model table
DECLARE @model VARBINARY(MAX) = (SELECT model_object FROM models WHERE
model_name = 'iris.dtree')

-- Run a prediction using the Iris_test data as input
-- Return all columns, including the probability for each species
EXEC sp_execute_external_script
    @language = N'R',
    @script = N'
            model = rxUnserializeModel(model);
            Iris_prediction = rxPredict(model, data=Iris_test)
            Iris_pred_results <- cbind(Iris_test, Iris_prediction)
```

```
        str(Iris_pred_results)
        ',
    @input_data_1 = N'
                SELECT
                    Sepal_Length,
                    Sepal_Width,
                    Petal_Length,
                    Petal_Width,
                    Species
                FROM Iris_test',
    @input_data_1_name = N'Iris_test',
    @output_data_1_name = N'Iris_pred_results',
    @params = N'@model varbinary(max)',
    @model = @model
WITH RESULT SETS (("Sepal_Length" FLOAT, "Sepal_Width" FLOAT,
"Petal_Length" FLOAT, "Petal_Width" FLOAT, Species VARCHAR(50),
setosa_Pred FLOAT, versicolor_Pred FLOAT, verginica_Pred FLOAT))
```

In the first part of the R script inside the `sp_execute_external_script` code, we have to unserialize our model again using `rxUnserializeModel`. With the model unserialized, we can perform a prediction of the input data. The last line of R code adds the probability columns for each Iris species to the input dataset. This means we end up with a single table as output that contains all the input columns as well as the columns generated by the scoring process.

We won't go into details about machine learning or machine learning algorithms in this book, but the problem we are trying to solve using machine learning in this case is one called classification. Machine learning algorithms can basically be grouped into three different categories: regression, classification, and clustering. With regression we are trying to predict a numerical value, for instance, the price of a car. Classification usually deals with predicting a categorical value, like the example we went through in this chapter: What species of Iris plant is this? Clustering algorithms try to predict a result by trying to group categories together based on their characteristics. In the Iris example we could also have chosen to use a clustering algorithm since there might be clear Iris species characteristics that tend to group together based on the species.

After running the code in Listing 7-8, we see the results shown in Figure 7-6. If you ran the code yourself, you might see some different results since we split our training and test data based on randomly selected rows.

Results Messages

| | Sepal_Length | Sepal_Width | Petal_Length | Petal_Width | Species | setosa_Pred | versicolor_Pred | verginica_Pred |
|---|---|---|---|---|---|---|---|---|
| 1 | 4.8 | 3.4 | 1.9 | 0.2 | setosa | 1 | 0 | 0 |
| 2 | 5.1 | 3.8 | 1.9 | 0.4 | setosa | 1 | 0 | 0 |
| 3 | 5.2 | 2.7 | 3.9 | 1.4 | versicolor | 0 | 0.916666666666667 | 0.0833333333333333 |
| 4 | 5.2 | 3.5 | 1.5 | 0.2 | setosa | 1 | 0 | 0 |
| 5 | 5.3 | 3.7 | 1.5 | 0.2 | setosa | 1 | 0 | 0 |
| 6 | 5.4 | 3.4 | 1.5 | 0.4 | setosa | 1 | 0 | 0 |
| 7 | 5.5 | 2.4 | 3.7 | 1 | versicolor | 0 | 0.916666666666667 | 0.0833333333333333 |
| 8 | 5.5 | 2.4 | 3.8 | 1.1 | versicolor | 0 | 0.916666666666667 | 0.0833333333333333 |
| 9 | 5.6 | 2.9 | 3.6 | 1.3 | versicolor | 0 | 0.916666666666667 | 0.0833333333333333 |
| 10 | 5.7 | 2.6 | 3.5 | 1 | versicolor | 0 | 0.916666666666667 | 0.0833333333333333 |
| 11 | 5.7 | 2.8 | 4.5 | 1.3 | versicolor | 0 | 0.916666666666667 | 0.0833333333333333 |
| 12 | 5.7 | 3 | 4.2 | 1.2 | versicolor | 0 | 0.916666666666667 | 0.0833333333333333 |
| 13 | 5.8 | 2.7 | 4.1 | 1 | versicolor | 0 | 0.916666666666667 | 0.0833333333333333 |
| 14 | 6 | 2.2 | 4 | 1 | versicolor | 0 | 0.916666666666667 | 0.0833333333333333 |

***Figure 7-6.*** *Scored results for the data inside the Iris_test table using our trained machine learning model*

Performing a prediction using the `sp_execute_external_script` method works perfectly fine and gives you maximum flexibility in terms of what you can do using R code. However, it does result in quite a lot of lines of code. Another method we have available inside SQL Server is using the `PREDICT` function; `PREDICT` is far easier to use, has a simpler syntax, and, in general, performs faster than `sp_execute_external_script`. It does have its drawbacks though, for instance, you cannot write custom R code to perform additional steps on the data and you are required to use a serialized model that was trained using a Revolution Analytics algorithm (by using `sp_execute_external_script` you can basically use every algorithm available in R or R libraries).

We performed the same scoring on our data inside the Iris_test table using the `PREDICT` function in the code of Listing 7-9.

***Listing 7-9.*** Running a model prediction using the PREDICT function

```
DECLARE @model VARBINARY(MAX) = (SELECT model_object FROM models WHERE
model_name = 'iris.dtree')
-- Alternative method is using the PREDICT function
SELECT
```

```
    Iris_test.*,
    pred.*
FROM PREDICT(MODEL = @model, DATA = dbo.Iris_test as Iris_test)
WITH(setosa_Pred FLOAT, versicolor_Pred FLOAT, virginica_Pred FLOAT) AS pred
```

As you can directly see, PREDICT is far more readable than sp_execute_external_script and, for those more familiar with T-SQL, far easier to understand. In a sense, we are joining the model, and its outputs, to the data inside the Iris_test table. We need to supply the column names and datatypes of the prediction output inside the WITH clause and can select what we want to return using the SELECT statement. In this case we are selecting all the columns of the Iris_test table together with all the columns that are returned by the prediction, and the results should look like those shown in Figure 7-7.

| | Sepal_Length | Sepal_Width | Petal_Length | Petal_Width | Species | setosa_Pred | versicolor_Pred | virginica_Pred |
|---|---|---|---|---|---|---|---|---|
| 1 | 4.8 | 3.4 | 1.9 | 0.2 | setosa | 1 | 0 | 0 |
| 2 | 5.1 | 3.8 | 1.9 | 0.4 | setosa | 1 | 0 | 0 |
| 3 | 5.2 | 2.7 | 3.9 | 1.4 | versicolor | 0 | 0.916666666666667 | 0.0833333333333333 |
| 4 | 5.2 | 3.5 | 1.5 | 0.2 | setosa | 1 | 0 | 0 |
| 5 | 5.3 | 3.7 | 1.5 | 0.2 | setosa | 1 | 0 | 0 |
| 6 | 5.4 | 3.4 | 1.5 | 0.4 | setosa | 1 | 0 | 0 |
| 7 | 5.5 | 2.4 | 3.7 | 1 | versicolor | 0 | 0.916666666666667 | 0.0833333333333333 |
| 8 | 5.5 | 2.4 | 3.8 | 1.1 | versicolor | 0 | 0.916666666666667 | 0.0833333333333333 |
| 9 | 5.6 | 2.9 | 3.6 | 1.3 | versicolor | 0 | 0.916666666666667 | 0.0833333333333333 |
| 10 | 5.7 | 2.6 | 3.5 | 1 | versicolor | 0 | 0.916666666666667 | 0.0833333333333333 |
| 11 | 5.7 | 2.8 | 4.5 | 1.3 | versicolor | 0 | 0.916666666666667 | 0.0833333333333333 |
| 12 | 5.7 | 3 | 4.2 | 1.2 | versicolor | 0 | 0.916666666666667 | 0.0833333333333333 |
| 13 | 5.8 | 2.7 | 4.1 | 1 | versicolor | 0 | 0.916666666666667 | 0.0833333333333333 |
| 14 | 6 | 2.2 | 4 | 1 | versicolor | 0 | 0.916666666666667 | 0.0833333333333333 |

***Figure 7-7.*** *Iris species prediction using PREDICT*

Now that we have trained a machine learning model, and scored data using it, inside SQL Server Machine Learning Services, you should have a general idea of the capabilities of these methods. In general, we believe In-Database Machine Learning Services is especially useful when all, or the largest part, of your data is stored inside SQL Server databases. With the model stored inside a SQL Server database as well, you can build solutions that are able to (near) real-time score data as soon as it is stored inside your SQL Server database (for instance, by using triggers that call the PREDICT function). If you want to, you are not limited to just SQL Server tables however. As you have seen in earlier chapters, we can map data stored inside the Spark cluster (or on other systems all together) using external tables and pass that data to the In-Database Machine Learning Services.

In some situations, however, you cannot use In-Database Machine Learning Services, perhaps because your data doesn't fit inside SQL Server, either by size or by data type, or you are more familiar with working on Spark. In any of those cases, we always have the option of performing machine learning tasks on the Spark portion of the Big Data Cluster which we are going to explore in more detail in the next section.

# Machine Learning in Spark

Since Big Data Clusters are made up from SQL Server and Spark nodes, we can easily choose to run our machine learning processes, from training to scoring, inside the Spark platform. There are many reasons we can come up with why you would choose Spark over SQL for a machine learning platform (and vice versa). However, when you have a very large dataset that doesn't make sense to load into a database, you are more or less stuck on using Spark since Spark can handle large datasets very well and can train various machine learning algorithms in the same distributed nature as it handles data processing.

As expected on an open, distributed, data processing platform, there are many libraries available which you can use to satisfy your machine learning needs. In this book we decided on using the built-in Spark ML libraries which provide a large selection of different algorithms and should cover most of your advanced analytical needs.

Just like we did for the In-Database Machine Learning Services for SQL Server section, we need to get some data inside Spark to work with. For the sake of simplicity, we decided on reusing the Iris dataset we also used for the SQL Server section. Just like we did in the previous chapter, all the data processing, wrangling, and analysis we are doing in Spark happen on a dataframe. Assuming you worked through the examples in the previous SQL Server section, we are going to extract the Iris dataset from inside the SQL Server Master Instance and load it into a dataframe in Spark using the code of Listing 7-10. If you are unfamiliar with connecting to the SQL Server Master Instance through Spark, we suggest reading the last section of the previous chapter where we go into detail how you can make this scenario work.

***Listing 7-10.*** Reading data from the SQL Server Master Instance

```
# Before we get started, let's get the Iris data from the database/table we
# created in the previous section
df_Iris = spark.read.format("jdbc") \
    .option("url", "jdbc:sqlserver://master-0.master-svc;databaseName=InDBML") \
```

```
    .option("dbtable", "dbo.Iris") \
    .option("user", "[username]") \
    .option("password", "[password]").load()
```

If we look at some of the contents of the df_Iris dataframe, using the `df_Iris.show(10)` command, we should see that all the Iris species characteristics, as well as the species itself, are present in the dataframe (Figure 7-8).

```
+------------+-----------+------------+-----------+-------+
|Sepal_Length|Sepal_Width|Petal_Length|Petal_Width|Species|
+------------+-----------+------------+-----------+-------+
|         5.1|        3.5|         1.4|        0.2| setosa|
|         4.9|        3.0|         1.4|        0.2| setosa|
|         4.7|        3.2|         1.3|        0.2| setosa|
|         4.6|        3.1|         1.5|        0.2| setosa|
|         5.0|        3.6|         1.4|        0.2| setosa|
|         5.4|        3.9|         1.7|        0.4| setosa|
|         4.6|        3.4|         1.4|        0.3| setosa|
|         5.0|        3.4|         1.5|        0.2| setosa|
|         4.4|        2.9|         1.4|        0.2| setosa|
|         4.9|        3.1|         1.5|        0.1| setosa|
+------------+-----------+------------+-----------+-------+
only showing top 10 rows
```

***Figure 7-8.*** *df_Iris dataframe top ten rows*

With our data inside a dataframe in Spark, we are almost ready to start to do some machine learning. First thing we need to handle though is the loading of a number of Spark ML libraries as shown in the code in Listing 7-11.

***Listing 7-11.*** Loading machine learning libraries

```
# To perform machine learning tasks, we need to import a number of libraries
# In this case we are going to perform classification
from pyspark.ml.classification import *
from pyspark.ml.evaluation import *
from pyspark.ml.feature import *
```

In this case, since we are doing a so-called classification problem, we only need to import the `pyspark.ml.classification` libraries together with the libraries we need to perform some modification to the features (which is another name for the columns of our dataframe in this case) of the dataframe and evaluate our model performance.

After the libraries are loaded, we are going to perform some modifications on our dataframe to make it suitable to work for our machine learning algorithm. Different machine learning algorithms have different requirements in terms of your data, for instance, some algorithms only work on numerical values as input, just like the classification algorithm we are using. The code of Listing 7-12 performs a number of tasks on our df_Iris dataframe.

***Listing 7-12.*** Process the data so it is suitable for machine learning

```
# We are going to combine all the features we need to predict the Iris species
# into a single vector feature
feature_cols = df_Iris.columns[:-1]
assembler = VectorAssembler(inputCols=feature_cols, outputCol="features")
df_Iris = assembler.transform(df_Iris)
df_Iris = df_Iris.select("features", "Species")

# Since we are going to perform logistic regression, we are going to convert
# the string values inside species to a numerical value
label_indexer = StringIndexer(inputCol="Species", outputCol="label").fit(df_Iris)
df_Iris = label_indexer.transform(df_Iris)
```

The first code section combines the different features inside a new column called "features." All of these features are Iris species characteristics and they are combined into a single format called a vector (we will take a look at how this visually looks a bit further down in the book). The line `feature_cols = df_Iris.columns[:-1]` selects all the columns of the dataframe except the rightmost column which is the actual species of the Iris plant.

In the second section, we are mapping the different Iris species to a numerical value. The algorithm we are going to use to predict the Iris species requires numerical input, which means we have to perform a conversion. This is not unusual in the realm of machine learning and data science. In many cases you have to convert a string value to a numerical value so the algorithm can work with it. After the conversion from string to numerical, we add a new column called "label" which contains the species in a numerical value.

In the next step, we are only selecting the features and the label column from the df_Iris dataframe and return the top ten rows (code of Listing 7-13 results in Figure 7-9) to give you an idea how the data looks after the transformations we've performed in the previous code segment.

***Listing 7-13.*** Only select the features and label dataframe columns

```
# We only need the feature column and the label column
df_Iris = df_Iris.select("features", "label")

df_Iris.show(10)
```

```
+-----------------+-----+
|         features|label|
+-----------------+-----+
|[5.1,3.5,1.4,0.2]|  2.0|
|[4.9,3.0,1.4,0.2]|  2.0|
|[4.7,3.2,1.3,0.2]|  2.0|
|[4.6,3.1,1.5,0.2]|  2.0|
|[5.0,3.6,1.4,0.2]|  2.0|
|[5.4,3.9,1.7,0.4]|  2.0|
|[4.6,3.4,1.4,0.3]|  2.0|
|[5.0,3.4,1.5,0.2]|  2.0|
|[4.4,2.9,1.4,0.2]|  2.0|
|[4.9,3.1,1.5,0.1]|  2.0|
+-----------------+-----+
only showing top 10 rows
```

***Figure 7-9.*** *modified df_Iris dataframe*

As you can see from Figure 7-9, all of the features (Petal_Length, Petal_Width, etc.) have been transformed inside a single vector inside a single column of our dataframe. The label column now returns a number for the species, 2.0 being Setosa, 1.0 virginica, and 0.0 versicolor.

Now that we have our entire dataframe converted into a format that is workable for our machine learning classification algorithm, we can split our data into a training dataframe and a testing dataframe like we did in the previous section as well. The code of Listing 7-14 handles the split in which 80% of the data goes into the `Iris_train` dataframe and the remaining 20% in the `Iris_test` dataframe.

***Listing 7-14.*** Split the dataframe into a training and testing dataframe

```
# Split the dataset
(Iris_train, Iris_test) = df_Iris.randomSplit([0.8, 0.2])
```

Now that we have our datasets ready for training, we can start the actual machine learning phase. The first thing we need to do is to initialize the machine learning algorithm (Listing 7-15). In this part we can supply which algorithm we want to use and various parameters (also called hyperparameters) we want to configure during the training phase of the machine learning model.

***Listing 7-15.*** Initiate the classifier

```
# Initiate the classifier, in this case LogisticRegression
lr = LogisticRegression(maxIter=10, tol=1E-6, fitIntercept=True)
```

In this case we have chosen to use a logistic regression algorithm to try and predict which species of Iris a plant belongs to, based on its characteristics. We are going to ignore the algorithm parameters for now. When you are in the phase when you try to optimize and tune your model, you will frequently go back to the parameters (either manually or programmatically) and modify them to find the optimal setting.

Training the model is actually very easy and straightforward and, in this case, can be achieved by a single line of PySpark code (Listing 7-16).

***Listing 7-16.*** Train the model

```
# Train the multiclass model
model = lr.fit(Iris_train)
```

After the preceding code (Listing 7-16) finished running, we have access to a trained machine learning model in the form of the variable "`model`." We can then use the trained model to perform predictions on our test dataset to analyze how well it performed. Using the code of Listing 7-17, we are going to "fit" the trained model on our test dataset and return the top 20 results which are shown in Figure 7-10.

***Listing 7-17.*** Perform a prediction

```
# Predict on our test dataset using the model we trained
# and return the predictions
Iris_pred = model.transform(Iris_test)

Iris_pred.show(20)
```

```
+-----------------+-----+--------------------+--------------------+----------+
|         features|label|       rawPrediction|         probability|prediction|
+-----------------+-----+--------------------+--------------------+----------+
|[4.6,3.2,1.4,0.2]|  2.0|[0.30991078639312...|[3.35040440695999...|       2.0|
|[4.6,3.4,1.4,0.3]|  2.0|[-0.3750826150434...|[4.04401905664896...|       2.0|
|[4.8,3.1,1.6,0.2]|  2.0|[1.19146977338786...|[7.12505508251146...|       2.0|
|[5.0,2.0,3.5,1.0]|  0.0|[6.30375932984991...|[0.87175896041068...|       0.0|
|[5.0,3.3,1.4,0.2]|  2.0|[0.83587934080054...|[7.43466366511768...|       2.0|
|[5.1,3.5,1.4,0.2]|  2.0|[0.37926677687117...|[8.64163143944124...|       2.0|
|[5.4,3.4,1.7,0.2]|  2.0|[1.53357534235435...|[4.32896208703599...|       2.0|
|[5.4,3.7,1.5,0.2]|  2.0|[0.41091088753656...|[4.04056483469234...|       2.0|
|[5.4,3.9,1.3,0.4]|  2.0|[-0.4014412200652...|[4.54625576899838...|       2.0|
|[5.5,2.4,3.7,1.0]|  0.0|[6.15154045507437...|[0.97112658101474...|       0.0|
|[5.6,2.9,3.6,1.3]|  0.0|[4.59088469450658...|[0.97386844668033...|       0.0|
|[5.6,3.0,4.1,1.3]|  0.0|[4.54103727639938...|[0.96102614316431...|       0.0|
|[5.7,2.8,4.5,1.3]|  0.0|[5.65763396206991...|[0.88562559392923...|       0.0|
|[5.8,2.7,5.1,1.9]|  1.0|[6.47541333408135...|[0.06033573156837...|       1.0|
|[5.9,3.0,4.2,1.5]|  0.0|[5.21905356429863...|[0.89724043667644...|       0.0|
|[6.1,2.6,5.6,1.4]|  1.0|[7.80857557133337...|[0.39206132823967...|       1.0|
|[6.2,2.8,4.8,1.8]|  1.0|[6.84252818274330...|[0.28829058044637...|       1.0|
|[6.3,2.5,4.9,1.5]|  0.0|[8.16207847548803...|[0.47150854117348...|       1.0|
|[6.3,2.9,5.6,1.8]|  1.0|[7.17991530788880...|[0.16034813924546...|       1.0|
|[6.4,2.8,5.6,2.1]|  1.0|[7.69286100981022...|[0.02379051709345...|       1.0|
+-----------------+-----+--------------------+--------------------+----------+
only showing top 20 rows
```

***Figure 7-10.*** *Prediction results on our test dataset*

As you can see in Figure 7-10, our model performed a good job on the test dataset. In the top 20 rows that were returned by the command, only a single row had a prediction for a different species instead of the actual one (we predicted virginica while it should have been versicolor). While we could analyze each and every row to look for differences between the actual species and the predicted species, a far faster way to look at model performance is by using the Spark ML evaluation library which we loaded earlier.

The code of Listing 7-18 evaluated the model performance against our test dataset and measured it on the performance metric accuracy. Accuracy is frequently used to measure how well a classification model is performing and is the ratio of correct predictions divided by the number of incorrect predictions.

***Listing 7-18.*** Measuring model performance

```
# How good did our model perform?
evaluator = MulticlassClassificationEvaluator(metricName='accuracy')
accuracy = evaluator.evaluate(Iris_pred)
print("Accuracy: " + format(accuracy))
```

The results the preceding code returns will probably vary each time you run the code. This is because the dataset we are using is rather small and we perform a randomize split, which means the number of unique species which ends up in the training and testing datasets has a huge influence on model performance. We ended up with the results shown in Figure 7-11, which is quite a respectable level of accuracy.

```
Accuracy: 0.935483870968
```

***Figure 7-11.*** *Accuracy of our trained model*

With our model trained and tested, we can take additional steps depending on what we are planning to do with the model. If we are interested in optimizing model performance more, we could go back and tune our algorithm parameters before training the model again. Perhaps it would also be useful, in this scenario, to look how good the split is between the training and test dataset since that has a huge impact on the model accuracy and there are a hundred more things we could do to optimize our model even further if we wanted to (even selecting a different algorithm to see if that predicts better than the current one).

Another thing we could do is store the model. We are way more flexible in that area than inside SQL Server In-Database Machine Learning Services where the model had to be serialized and stored inside a table. In the case of Spark, we can choose different methods and libraries to store our models. For instance, we can use a library called Pickle to store our model on the filesystem, or use the `.save` function on the model variable to store it on an HDFS location of our choosing. Whenever we need our trained model to score new data, we can simply load it from the filesystem and use it to score the new data.

# Summary

In this chapter we explored the various methods available to perform machine learning tasks inside SQL Server Big Data Clusters. We looked at SQL Server In-Database Machine Learning Services which allowed us to train, utilize, and store machine learning models directly inside the SQL Server Master Instance using a combination of T-SQL queries and the new `sp_execute_external_script` procedure. In the Spark department, we also have a wide variety of machine learning capabilities available to use. We used the Spark ML library to train a model on a dataframe and used it to score new data. Both of the methods have their strengths and weaknesses, but having both of these solutions available inside a single box allows optimal flexibility for all our machine learning needs.

# CHAPTER 8

# Create and Consume Big Data Cluster Apps

One of the capabilities of SQL Server Big Data Clusters is the ability to build and run custom applications on its surface. This is actually a very powerful feature, since it allows you to script and run a wide variety of solutions on top of your Big Data Cluster. For instance, you can create an application, or app as we will call it in the remainder of this chapter, to perform various maintenance tasks on top of your data like a database backup. Another example is the ability to create an entry point for your machine learning processes through a REST API, a use case which we will explore later in this chapter.

Apps that you create on your Big Data Cluster can, at the moment of writing this book, be written in R and Python, and there is an additional option to run SQL Server Integration Services (SSIS) packages as well. By creating apps, you can utilize all the computational resources available inside the Big Data Cluster as well as access all the data that is stored inside of it.

Apps inside Big Data Clusters are run inside a dedicated container and can be replicated and scaled across the cluster. This means that you can make your apps handle parallel workloads and be high-performant solutions.

In this chapter we are going to create an app that will use a pretrained machine learning model to classify the species of Iris plants, much like we have done in the previous chapter which focused on developing machine learning solutions inside Big Data Clusters. By building an app to score data using a machine learning solution, we can easily operationalize that model through a REST API. This means that applications that you use or build yourself can receive a prediction directly from the Big Data Cluster through JSON messages, allowing near real-time scoring directly from your application without the need to store and process the data first inside the Big Data Cluster.




B. Weissman and E. van de Laar, *SQL Server Big Data Clusters*,
https://doi.org/10.1007/978-1-4842-5985-6_8

## Create a Big Data Cluster App

There are two methods we can use to deploy apps to the Big Data Cluster, Visual Studio Code through the App Deploy Extension and through the azdata command-line utility. We are going to focus on the latter method to create and deploy our app.

Before we can deploy our app, we first have to write it. As mentioned in the introduction of this chapter, apps can be written in R or Python and we have selected R as the language of choice for our app. While it is not strictly necessary to have access to R if you want to follow the examples in this chapter, it can be useful if you want to train the machine learning model we are using inside our app yourself. In any case, the pretrained model and the other files that are required for app deployment are available for download at this book's GitHub page.

Since we are going to create a Big Data Cluster app that is going to score new data using a machine learning model, we need to create and store the model first. The code in Listing 8-1 will use the built-in Iris dataset to create a machine learning model through a decision tree and store it inside an .RDS file (make sure to set a directory path before running the code). We will use the model stored inside the .RDS file later to score new data. You can execute the following code from an R session on your local computer. You can download and install R from `www.r-project.org/`.

***Listing 8-1.*** Building a prediction model in R

```
# Read the Iris data into a new dataframe
Iris_Data <- iris

# Change the column names
colnames(Iris_Data) <- c('Sepal_Length', 'Sepal_Width', 'Petal_Length',
'Petal_Width', 'Species')

# Sample a number of rows for splitting training and testing datasets
sample_size <- floor(0.75 * nrow(Iris_Data))

set.seed(1234)
train_id <- sample(seq_len(nrow(Iris_Data)), size = sample_size)

Iris_train <- Iris_Data[train_id, ]
Iris_test <- Iris_Data[-train_id, ]
```

```
# Train the model, a decision tree, on the training data
Iris_Dtree <- rpart(Species~., data = Iris_train, method = 'class')

# Save the model to disk
saveRDS(Iris_Dtree, "[folder path]/iris_dtree.rds")
```

As you can see in Listing 8-1, we went through the additional steps of splitting our data into a training and testing dataset. However, in the preceding code, we only use the training dataset to train the model and do not test its accuracy using the test dataset. We aren't necessarily interested in the model performance in this chapter, but rather the ability to use the pretrained model to score new data through our Big Data Cluster app. If you want to see how the model training performed, you can run the lines of the code in Listing 8-2, which will perform an Iris species prediction based on the model we trained and combine those predictions with the original testing dataset.

***Listing 8-2.*** Predict using our trained model

```
Iris_Predict <- predict(Iris_Dtree, Iris_test, method = 'class')
Prediction_results <- cbind(Iris_test, Iris_Predict)
```

Now that we have a pretrained model available to us inside and RDS file, we can take a look at the actual code required to create a Big Data Cluster app.

A Big Data Cluster app consists a minimum of two files: the actual code we are going to run inside the app and a YAML file that holds the configuration of our app. Both of these files, and any additional files you want to upload to the app container like our pretrained machine learning model, must all be stored inside a single directory as shown in Figure 8-1.

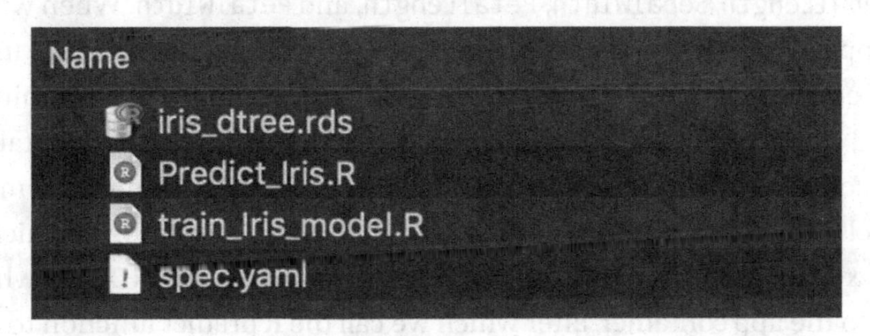


***Figure 8-1.*** *App files*

With the exception of the "spec.yaml" file, you are free to name your files in whichever way you want.

Let's first look at the contents of the "Predict_Iris.R" file. This file will contain the code necessary to load the pretrained model from the "iris_dtree.rds" file and perform a prediction based on the input variables we pass to the script file. The contents of the file can be seen in Listing 8-3.

***Listing 8-3.*** Contents of the Predict_Iris.R file

```
library(rpart)

runpredict <- function(SepalLength, SepalWidth, PetalLength, PetalWidth) {

    input_dataframe = data.frame(Sepal_Length = SepalLength, Sepal_Width =
    SepalWidth, Petal_Length = PetalLength, Petal_Width = PetalWidth)

    Iris_Dtree <- readRDS("iris_dtree.rds")

    Iris_Predict <- predict(Iris_Dtree, input_dataframe, method = 'class')

    result <- as.data.frame(Iris_Predict)
}
```

In the preceding code, we firstly load the R library needed to perform a prediction based on a decision tree. We also used the rpart library to train the model in the first place; hence, it is also required to load the library when we want to perform a prediction.

The entire processing through our script file is handled through an R function. This is necessary since we are going to define an entry point inside the spec.yaml file which is called whenever we run the app. In the function definition, I am defining four input variables, `SepalLength`, `SepalWidth`, `PetalLength`, and `PetalWidth`. When we are going to call our app, we are supplying these variables as input parameters for the model to perform a prediction. In the first line of code inside the function, I am grouping the input variables and storing them inside an R dataframe called `input_dataframe`, taking care to rename the columns to the identical format we also used when training the model. This is required, else the prediction would not know which data is residing in which column.

In the next step, we are loading the pretrained model from the RDS file which we also upload to the app container, after which we call the R predict function to perform a prediction using the model and the input dataframe. Finally, we convert the result of the prediction into a dataframe format and map it to the `result` variable.

Now that we have actually completed our application script, we have to create the `spec.yaml` file. For the example app we are deploying to our Big Data Cluster inside this chapter, the `spec.yaml` file looks like Listing 8-4.

***Listing 8-4.*** Contents of the spec.yaml file

```
name: predictiris
version: v1
runtime: R
src: ./Predict_Iris.R
entrypoint: runpredict
replicas: 1
poolsize: 1
inputs:
  SepalLength: numeric
  SepalWidth: numeric
  PetalLength: numeric
  PetalWidth: numeric
output:
  out: data.frame
```

Most of the contents of the `spec.yaml` file are pretty much self-explanatory. We supply a name and a version of the app, the runtime language, and the file that is called whenever we run our app. In the bottom section, we define our input parameters (which are identical to the ones we defined in the R function) and their datatypes, as well as the datatype of our output parameter. In this case we didn't explicitly set an output parameter name. This is because R automatically uses the last set variable (in our case `result`) as output when you call a function.

The more interesting parameters of the YAML file are the `replicas` and `poolsize` parameters. These are the parameters we can configure to replicate and scale our app. The input of the `replicas` parameter dictates how many pods should be deployed for the application and the `poolsize` configures how many occurrences of the app should be present inside a pod. The number of parallel operations your app is able to perform is the product of the calculation *replicas x poolsize*. For instance, having `replicas` configured to a value of 4 and the `poolsize` set to 2 will result in your app being able to handle 8 parallel requests. In our `spec.yaml`, we configured both these settings to be 1, meaning we will be able to handle one single request at a time.

With all the files we require for our app deployment ready and stored inside a single folder, we are ready to deploy the app to the Big Data Cluster. As we mentioned earlier, we are using the `azdata` program to perform the deployment.

Before we can connect to the Big Data Cluster, we need to retrieve the external IP of the `controller-svc-external` service. To do that, you can run the following command: `kubectl get svc controller-svc-external -n [clustername]`, where `[clustername]` is the name of your cluster. If you are using AKS to host your Big Data Cluster, you will first need to log on to Azure using the `az login` command.

Now that we have the IP and port number of the management service, we need to connect to it through azdata using our admin username and password. You supplied both of these during the deployment of your Big Data Clusters. The code of Listing 8-5 sets up a connection to your Big Data Cluster. Make sure to change the variables between [ ] to the values you have for your cluster.

***Listing 8-5.*** Login to the controller endpoint

```
azdata login --controller-endpoint https://[IP address]:30080 --controller-
username [username]
```

You will be asked to enter the password of the admin user. If everything went well, you should get a logged in successfully message.

Now that we are connected through azdata to our Big Data Cluster, we can deploy our application. To do that, we can use the code shown in Listing 8-6. Make sure to change the `[directory path]` to the path of the directory that holds your application files like the spec.yaml.

***Listing 8-6.*** Deploy Big Data Cluster app

```
azdata app create --spec [directory path]
```

In our case we called the following command (Listing 8-7).

***Listing 8-7.*** Deploying our Big Data Cluster app

```
azdata app create --spec /Users/enricovandelaar/Documents/BDC.
```

When running the preceding command, a number of validations will occur to check if the spec.yaml file is present and the input is correct. If everything is correct, you should receive the message shown in Figure 8-2.

```
PS C:\Users\Administrator> azdata app create --spec C:\users\Administrator\bdc_app
Application `predictiris/v1` successfully created.
```

***Figure 8-2.*** *Big Data Cluster app created*

Even though you receive a message that the app was created successfully, it isn't directly available. It usually takes around a minute before you can actually run your application after you create it. If you want to know the status of application creation, you can run the command in Listing 8-8.

***Listing 8-8.*** Retrieving App status through azdata

```
azdata app list -n predictiris
```

This returns the current status of the app deployment as shown in Figure 8-3.

```
App                                                    Name         State              Version
-----------------------------------------------------  -----------  -----------------  ---------
https://[illegible]/api/v1/app/predictiris/v1          predictiris  WaitingForCreate   v1
```

***Figure 8-3.*** *App creation status*

After a minute or so, we ran the command again and received a state of "Ready," shown in Figure 8-4, meaning we can continue with the next step to test our app.

```
App                                                    Name         State    Version
-----------------------------------------------------  -----------  -------  ---------
https://[illegible]/api/v1/app/predictiris/v1          predictiris  Ready    v1
```

***Figure 8-4.*** *App deployment completed and app is in the ready state*

When the app is in the "Ready" state, we can test it's functionality through the azdata program. If we defined any parameters, we need to supply them when calling the app, together with the name and version of the app which we supplied in the spec.yaml file. The command in Listing 8-9 calls our predictiris app together with a number of input parameters which we defined in the R script and YAML file.

***Listing 8-9.*** Run the app through azdata

```
azdata app run -n predictiris -v v1 --inputs PetalLength=1.4,PetalWidth=0.2,
SepalLength=5.1,SepalWidth=3.5
```

If everything completed successfully, we should get the results, in a JSON format, shown in Figure 8-5.

```
{
  "changedFiles": [],
  "consoleOutput": "",
  "errorMessage": "",
  "outputFiles": {},
  "outputParameters": {
    "out": {
      "setosa": [
        0.0
      ],
      "versicolor": [
        0.02702702702702702
      ],
      "virginica": [
        0.9729729729729729
      ]
    }
  },
  "success": true
}
```


***Figure 8-5.*** *App prediction results*

Since we are returning the output from the R script file inside a dataframe, the output is automatically converted to a JSON array. The prediction in the case of the predictiris app returns three output parameters which contain the probability for each possible Iris species. In this case, the virginica species seems to be the most likely giving the values of the input parameters we supplied with a certainty of 0.97 or 97%.

Whenever there is an issue with your app, you can in most cases see the error inside the "errorMessage" or "consoleOutput" sections of the returned JSON. In our case the app was executed successfully and we didn't run into any errors.

Now that we have our app deployed and tested, we can keep using the azdata method to call the app programmatically or on demand whenever we need to. Another method to execute the app, which I find far more elegantly, is through the REST API that is automatically created when we deploy our app.

# Consume Big Data Cluster Apps Through REST API

When we deploy our app, a dedicated container is created that holds our app and all additional files we supplied through the application folder. During the deployment, a RESTful web service is also created inside the container as an additional method to call the app. RESTful APIs use HTTP requests to perform tasks. In our case, we can use the REST API to call the app we created and return the outputs inside a JSON message. This can be very useful in situations where you create apps on your Big Data Cluster that you want to directly access from, for instance, your applications. Since all the code, and the data, resides on the Big Data Cluster, your application only needs to be able to send REST API calls and process the return messages returning data immediately into your application.

To make use of the REST API for our app, we need to perform a number of steps. The most important one being that we need to generate a token to securely call the REST API. A number of these steps need to be performed through a tool that can send REST API calls and process their results. In our case we used Postman (`www.getpostman.com/`) as the tool of our choice.

The first thing we need to do before we are able to connect to the REST API that belongs to our app is to generate a so-called "bearer token." Only by supplying this token in our REST API call can we access the app.

To generate a bearer token, we need to connect to the token URL. You can find the URL and port number you need to connect to by running the command in Listing 8-10.

***Listing 8-10.*** Retrieve app URL and port number through azdata

```
azdata app describe --name predictiris -v v1
```

Running the preceding command returns information about your app, in our case the `predictiris` app which is shown in Figure 8-6. The line we are after is returned in the "links" section and is the URL and port number of the "swagger" property.

```
{
  "input_param_defs": [
    {
      "name": "PetalLength",
      "type": "numeric"
    },
    {
      "name": "PetalWidth",
      "type": "numeric"
    },
    {
      "name": "SepalLength",
      "type": "numeric"
    },
    {
      "name": "SepalWidth",
      "type": "numeric"
    }
  ],
  "internal_name": "app1",
  "links": {
    "app": "https://[illegible]/api/v1/app/predictiris/v1",
    "swagger": "https://[illegible])/api/v1/app/predictiris/v1/swagger.json"
  },
  "name": "predictiris",
  "output_param_defs": [
    {
      "name": "out",
      "type": "data.frame"
    }
  ],
  "state": "Ready",
  "version": "v1"
}
```

***Figure 8-6.*** *Output of the app describe command for the predictiris app*

Copy the URL and port number, or write them down, for now and start Postman (or any other REST API call app you prefer). When Postman is started, we must change a setting to avoid an error. Since the Big Data Cluster is configuring self-signed certificates on its endpoint, we can potentially run into a security issue when we perform the REST API calls later on. Inside Postman you can find the SSL certificate verification inside the Preferences menu item as shown in Figure 8-7. Make sure to disable this setting before performing the REST API calls to your Big Data Cluster.

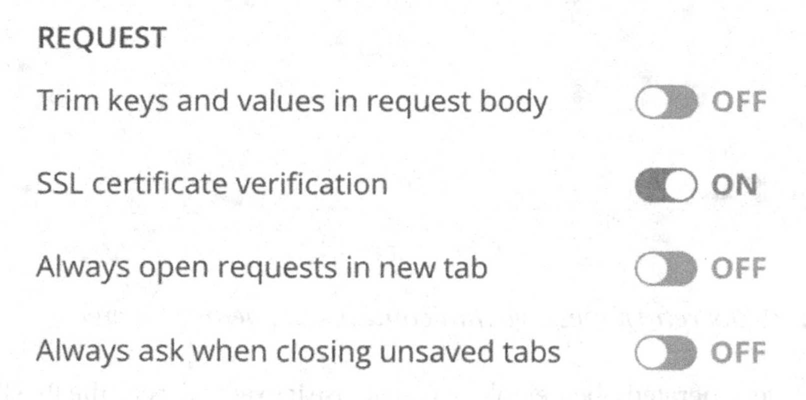


***Figure 8-7.*** *Request options inside Postman*

With the setting disabled, you can open a new tab inside Postman. Paste or enter the URL and port string we received from the app describe command into the request URL field and expand the URL with `/api/v1/token` and change the method to POST. Finally open the Authorization tab, change the Type to "Basic Auth," and enter your Big Data Cluster administrator username and password in the correct fields. Figure 8-8 shows a screenshot of Postman with all these items filled in for our Big Data Cluster and app URL.

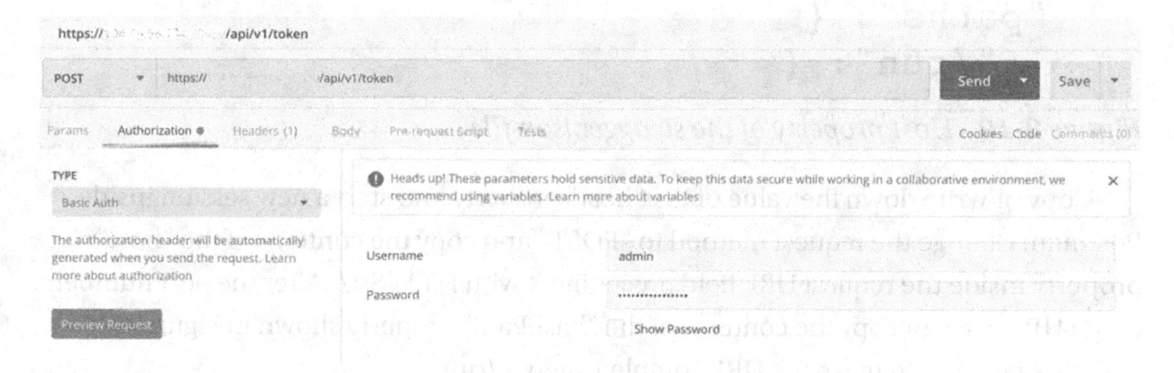


***Figure 8-8.*** *Postman setting to generate the bearer token*

With everything configured in Postman, click the Send button to send the request to the URL. If everything processed correctly, you should receive a return message that contains the bearer token inside the "access_token" property of the JSON response as shown in Figure 8-9 (we've removed the contents of the access_token and token_id properties in Figure 8-9).

***Figure 8-9.*** *JSON return message that contains the bearer token*

Now that we generated a bearer token, we can use it to actually call the REST API of the app itself. The URL of the app REST API is hidden by default and can be found in the swagger.json file which we can open by visiting the URL inside the "swagger" property that we received when running the `azdata app describe --name predictiris -v v1` command.

When you open the URL (in our case `https://104.46.56.134:30080/docs/swagger.json`), you can find a property in the JSON file called "host" as shown in Figure 8-10.

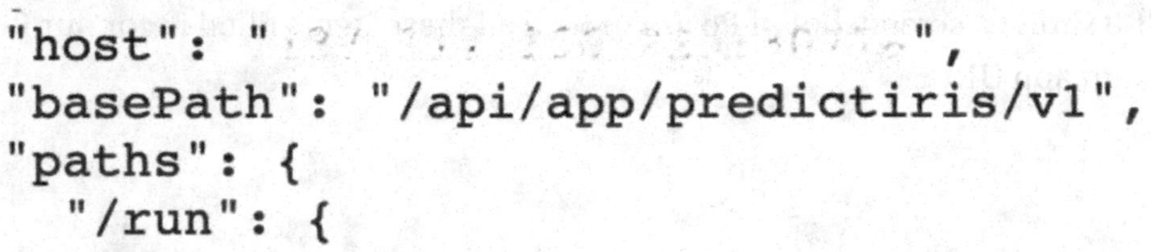

```
"host": "[illegible]",
"basePath": "/api/app/predictiris/v1",
"paths": {
  "/run": {
```

***Figure 8-10.*** *Host property of the swagger.json file*

Copy or write down the value of the "host" property and start a new session inside Postman. Change the request method to "POST" and copy the contents of the "host" property inside the request URL field preceding it with HTTPS://. After the port number of the URL, we can copy the contents of the "basePath" property shown in Figure 8-10 and, as a last step to make the URL complete, add a /run.

Go to the Authorization tab and this time select the option "Bearer Token" and add the token we received in the previous step inside the token field.

We now have one step left, generate the body content of our REST API call and supply the input parameters needed to perform the Iris species prediction. Inside Postman, click the Body tab, check the option "raw," and from the drop-down button, select "JSON (application/json)." Copy the contents of the code section in Listing 8-11 inside the body textarea to supply the input parameters needed for the predictiris app.

***Listing 8-11.*** Input parameters for predictiris app (JSON)

```
{
  "PetalLength": 1.4,
  "PetalWidth": 0.2,
  "SepalLength": 5.1,
  "SepalWidth": 3.5
}
```

With all of these areas filled in, the Postman screen should look like Figure 8-11.

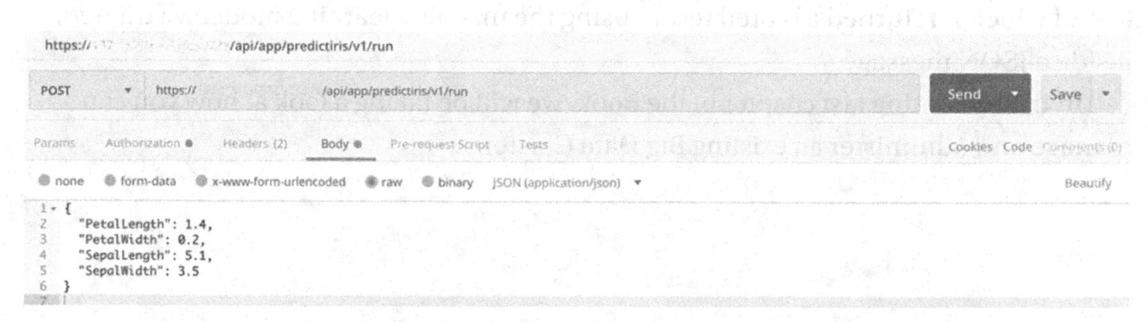


***Figure 8-11.*** *Body of the REST API call to the predictiris app*

Now all that is left to do is to click the "Send" button to send the JSON message to the predictiris Big Data Cluster app.

If everything was configured correctly, we should receive a return message that resembles the same output as when we executed the predictiris app using azdata containing the predicted probabilities for each species of Iris plant. The return message we received can be seen in Figure 8-12.

***Figure 8-12.*** *REST API response body with the probabilities of each Iris plant species*

# Summary

In this chapter, we took a look at creating and accessing Big Data Cluster applications. Big Data Cluster apps are a method to run containerized custom code inside the Big Data Cluster, for instance, to serve as an access point to perform machine learning scoring on a model that is stored inside the Big Data Cluster. We have created our own app that was able to predict the species of an Iris plant, uploaded it to the Big Data Cluster, and used azdata to execute the app. Apps are not only accessible through azdata though; by using a RESTful web service, we were able to access the app and send data to it of which it returned a scored result, using the machine learning model we trained, inside a JSON message.

In the upcoming last chapter of the book, we will be taking a look at how you can manage and administer an existing Big Data Cluster.

CHAPTER 9

# Maintenance of Big Data Clusters

Last but not least, we want to look at how you can check the health of your Big Data Cluster, how an existing Big Data Cluster can be upgraded to a newer version, and how you can remove a Big Data Cluster instance, if it's no longer needed.

## Checking the Status of a Big Data Cluster

Big Data Clusters provide you with two different portals from which to learn more about their current state and health. These portals provide metrics and insights on the status of the nodes as well as relating to log files. In addition to show, *azdata* can also provide you a high-level overview of your cluster's health.

## Retrieving a Big Data Cluster's Status Using azdata

To check your cluster's status from the command line, log in to your cluster using the command *azdata login*. As you can see in Figure 9-1, azdata will ask you for your namespace (your cluster's name), username, and password. Use the values provided during deployment.




B. Weissman and E. van de Laar, *SQL Server Big Data Clusters*,
https://doi.org/10.1007/978-1-4842-5985-6_9

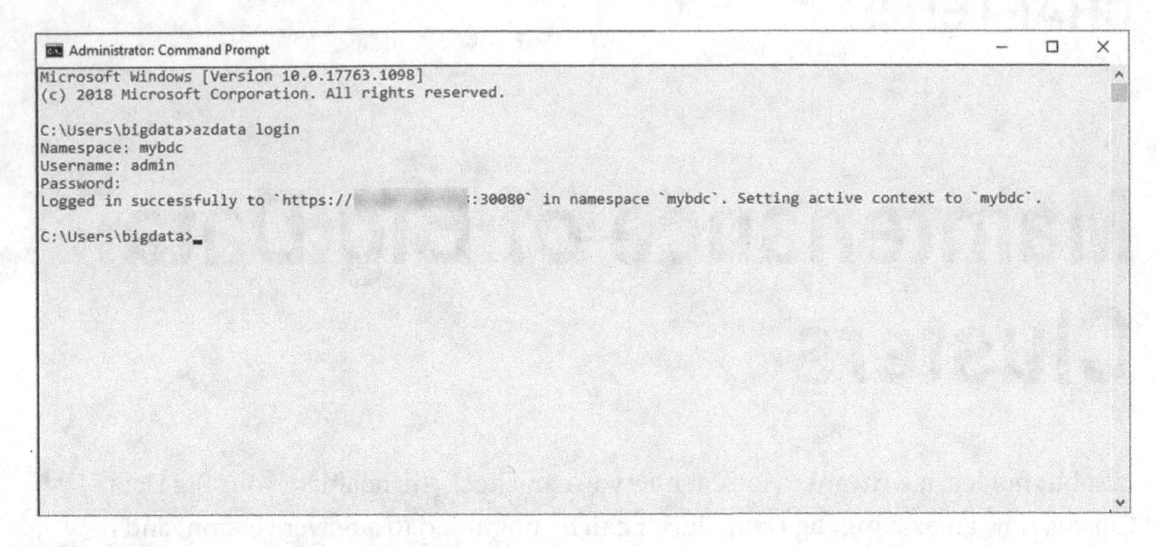


***Figure 9-1.*** *Output of azdata login*

Once you've successfully logged in, you can run the command *azdata bdc status show*. This will give you an overview of all your services, hopefully reporting them all as "healthy." A sample output is illustrated in Figure 9-2.

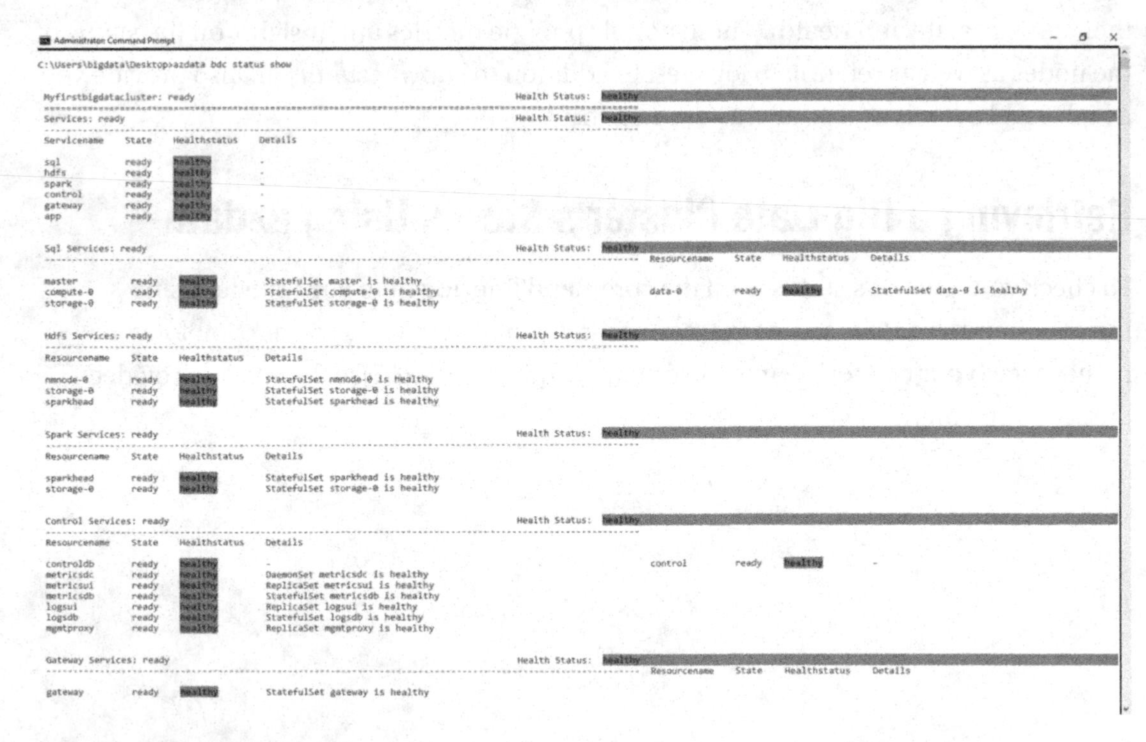


***Figure 9-2.*** *Output of azdata bdc status show*

After this, you can log out of the cluster by using *azdata logout*.

## Manage a Big Data Cluster Using ADS

Azure Data Studio gives you a more extensive view of your Big Data Cluster's status and layout. First, you need to connect to the cluster's controller endpoint, which you were provided at deployment.

To do so, look for the Big Data Clusters section in your ADS connections and click the "+" symbol as pointed out in Figure 9-3.

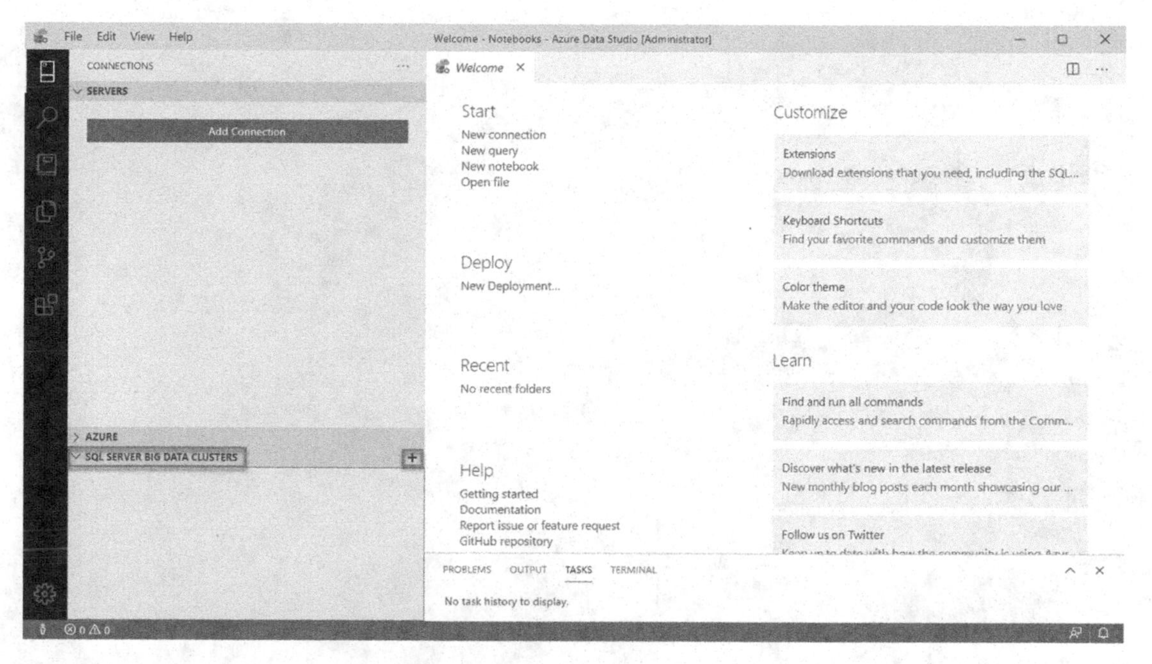


***Figure 9-3.*** *Big Data Clusters Connections in ADS*

In the next step, provide the endpoint URL as well as your credentials to log in to the cluster as shown in Figure 9-4.

Add New Controller

Cluster Management URL *

https:// :30080

Authentication type *

Basic

Username

admin

Password

••••••••

Remember Password

Add Cancel

*Figure 9-4. Add new Big Data Clusters Connection in ADS*

This will take you to your Big Data Cluster overview, which will show you the state and health status of every service as well as your endpoints as shown in Figure 9-5.

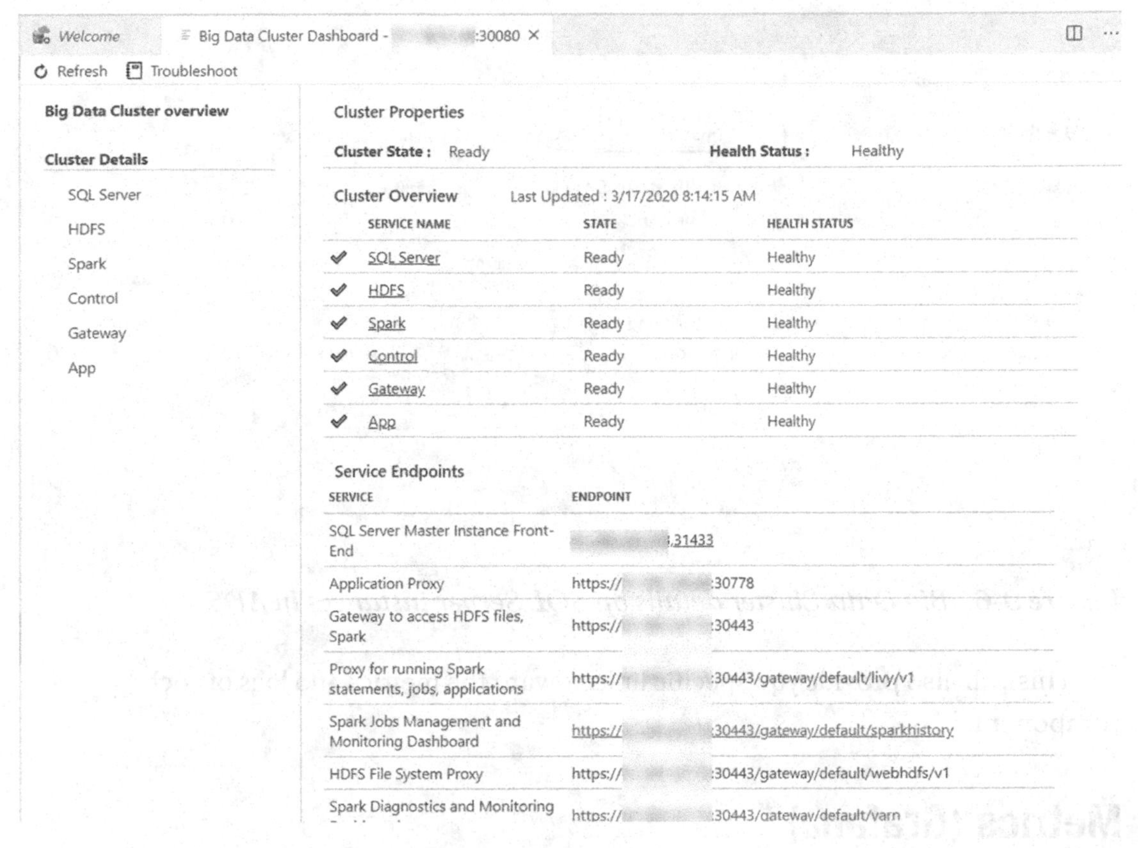


***Figure 9-5.*** *Big Data Cluster overview in ADS*

While the endpoints are more to be used as a reference, the overview itself can be very useful to retrieve more details about every single service and instance within your cluster.

If you click your *SQL Server* Service, for example, this will take you to an overview of all your SQL instances (master, compute, data, and storage) as shown in Figure 9-6.

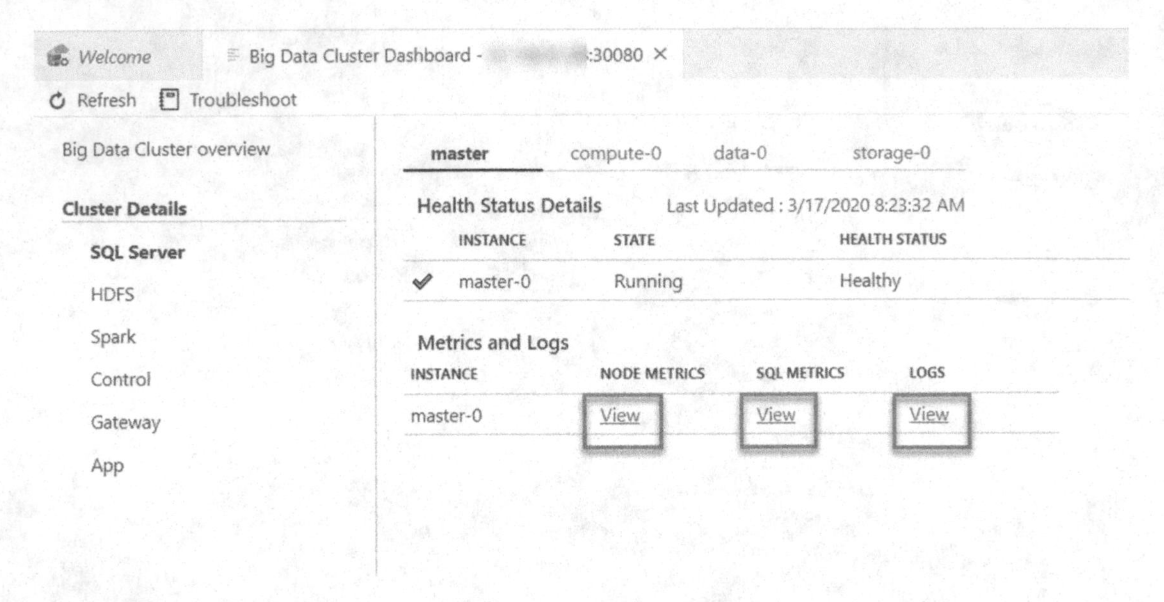


***Figure 9-6.*** *Big Data Cluster details on SQL Server instances in ADS*

This will also provide you specific links toward the metrics and logs of each component.

# Metrics (Grafana)

The Grafana Portal provides metrics and insights on the status on the node itself as well as more SQL-specific metrics where applicable. The credentials to log in to the portal will be the same ones you also used to connect to your cluster in Azure Data Studio.

## Node Metrics

Node metrics are typical performance indicators like CPU, RAM, and disk usage as shown in Figure 9-7.

***Figure 9-7.** Grafana Portal – node metrics*

In addition to the "big picture," you can also get detailed information for every single component like a specific disk or network interface.

When running into performance issues, this is always a good starting point. Obviously, this can also be a great indicator whether you overprovisioned your cluster.

## SQL Metrics

While the node metrics were focused on the physical side of the node, the SQL metrics as shown in Figure 9-8 provide information like wait time or number of waiting tasks by wait type, transactions, and requests per second and other valuable metrics to understand more about the status of the SQL components within the cluster.

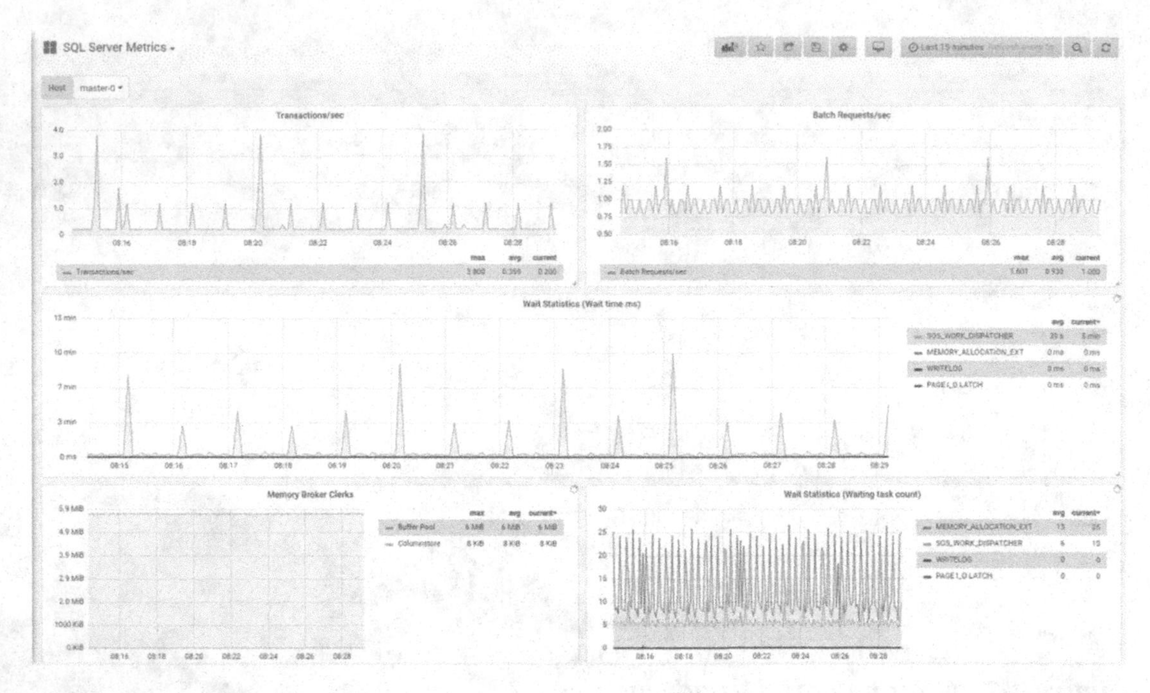


*Figure 9-8.* *Grafana Portal – SQL metrics*

Except for the master instance, which could also be reached through SSMS or Azure Data Studio, you usually don't connect to any of the other nodes directly, so think of these metrics as your replacement for activity monitor.

## Log Search Analytics (Kibana)

The Kibana dashboard as shown in Figure 9-9 on the other hand provides you an insight into the log files of the selected pod/node.

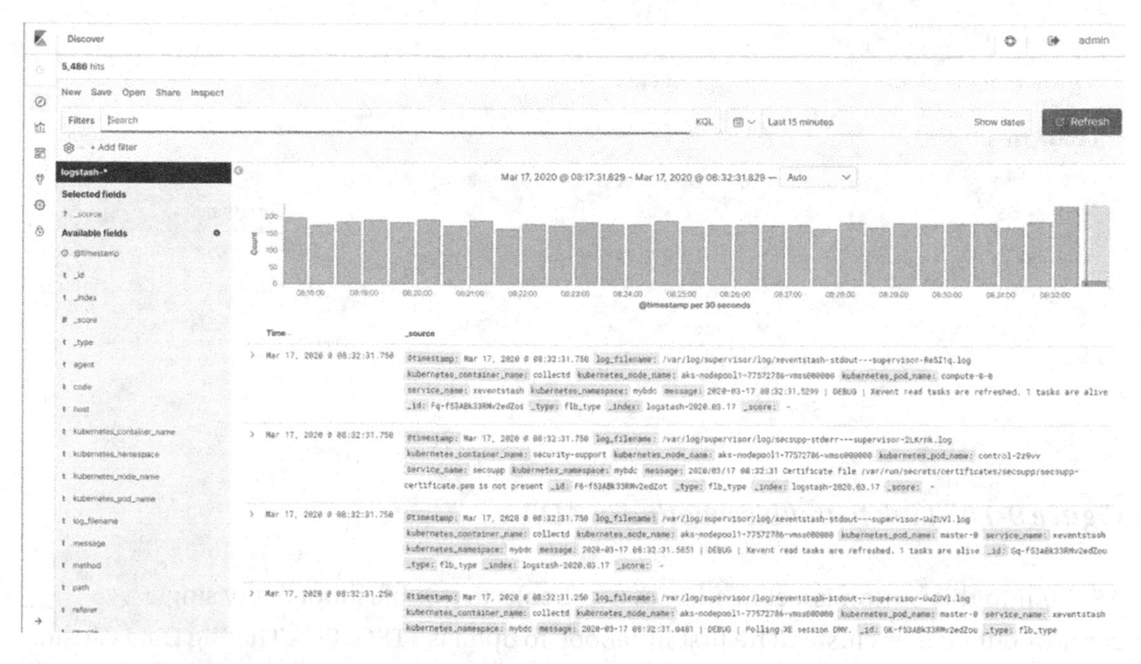


***Figure 9-9.** Kibana Portal - overview*

Kibana is part of the elastic stack. It also provides options to create visualizations and dashboard on top of your log files. If you want to learn more about it, its website `www.elastic.co/products/kibana` is a great starting point!

# Troubleshooting Big Data Clusters

At some point, your Big Data Cluster will probably run into a problem - from insufficient disk space to a faulty component. Azure Data Studio also provides guidance and tools on how to find and potentially fix the cause of such an error.

If you navigate back to the Big Data Cluster overview, you will see a button "Troubleshoot" as pointed out in Figure 9-10.

Refresh Troubleshoot

**Big Data Cluster overview**

**Cluster Details**

SQL Server
HDFS
Spark
Control
Gateway
App

Cluster Properties

**Cluster State :** Ready **Health Status :** Healthy

Cluster Overview Last Updated : 3/17/2020 8:31:56 AM

| SERVICE NAME | STATE | HEALTH STATUS |
|---|---|---|
| SQL Server | Ready | Healthy |
| HDFS | Ready | Healthy |
| Spark | Ready | Healthy |
| Control | Ready | Healthy |
| Gateway | Ready | Healthy |
| App | Ready | Healthy |

***Figure 9-10.** Link to troubleshooting in ADS*

Behind this button is a collection of notebooks to troubleshoot every single component of your cluster. The first notebook to open is "TSG100 – The Big Data Cluster troubleshooter" which will guide you through a full debugging of your Big Data Cluster. If you have already narrowed down which service is causing issues, you can also navigate directly to the analyzer notebook for that specific component on the left as shown in Figure 9-11.

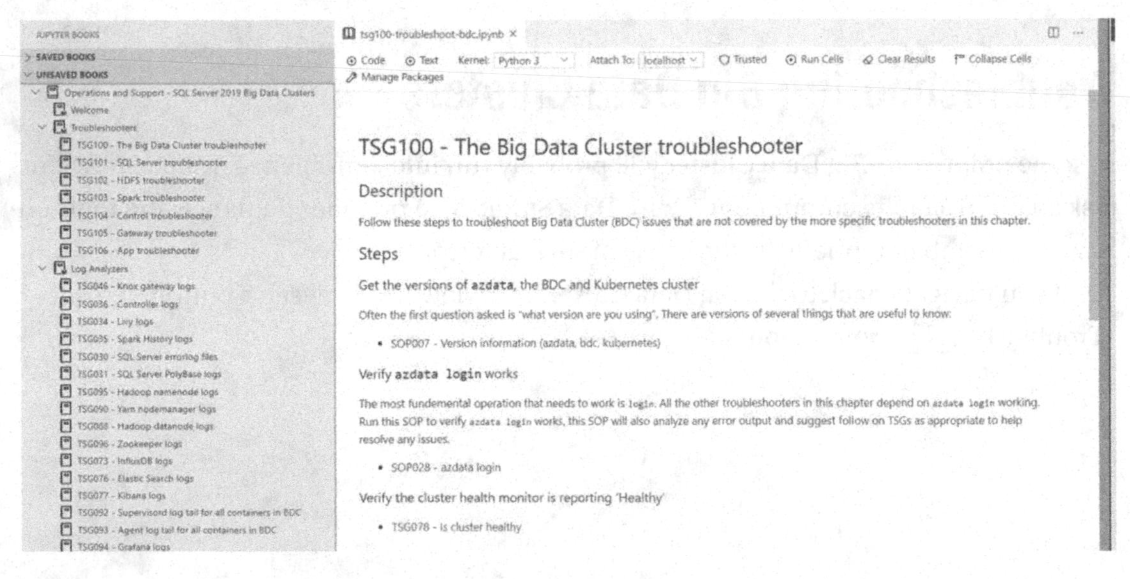


***Figure 9-11.** Troubleshooting in ADS*

The notebooks are grouped by category as illustrated in Figure 9-12 and are always your first starting point when you're experiencing problems with a Big Data Cluster.

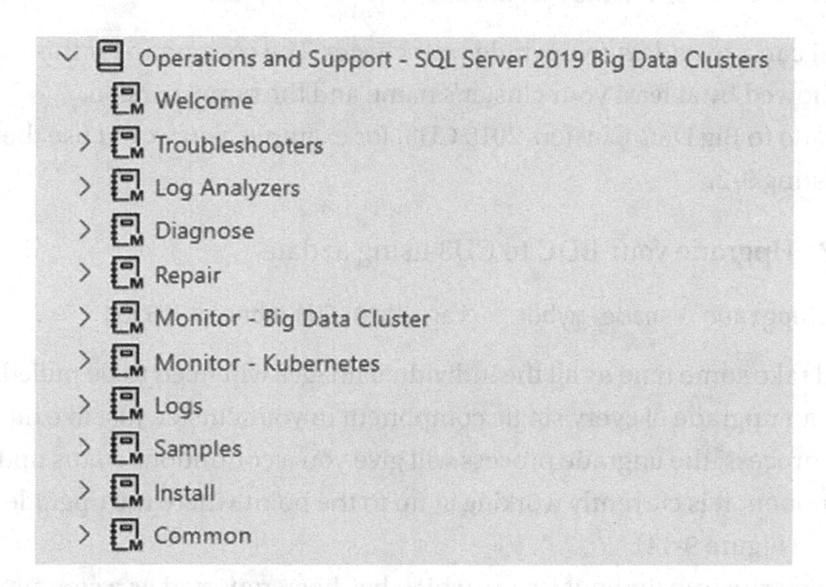


***Figure 9-12.*** *Troubleshooting categories in ADS*

# Upgrading Big Data Clusters

Just like any other version of SQL Server, Big Data Clusters receive regular cumulative updates (CU) during the version's maintenance time frame. To check your installation's version, you can just run *SELECT @@VERSION* in either SQL Server Management Studio or Azure Data Studio. Let's assume your current version is CU1 as shown in Figure 9-13.

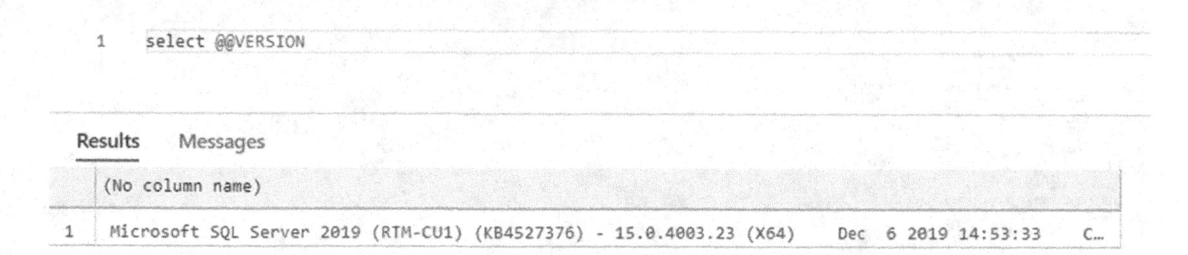


***Figure 9-13.*** *Output of SELECT @@VERSION*

If you want to upgrade your Big Data Cluster to a new version, first make sure that you have the latest version of azdata installed. To do so, run the code in Listing 9-1 just like when you first installed azdata.

***Listing 9-1.*** Update azdata to the latest version

```
pip3 install -r https://aka.ms/azdata
```

Now you can use *azdata* to upgrade your cluster. The command for this is *azdata bdc upgrade*, followed by at least your cluster's name and the target version.

To upgrade to Big Data Clusters 2019 CU3, for example, you would use the command shown in Listing 9-2.

***Listing 9-2.*** Upgrade your BDC to CU3 using azdata

```
azdata bdc upgrade --name mybdc --tag 2019-CU3-ubuntu-16.04
```

This will take some time as all the individual images will need to be pulled first followed by an upgrade of every single component in your cluster. Just like during the installation process, the upgrade process will give you a continuous status update on which component it is currently working at up to the point where the upgrade process is complete (see Figure 9-14).

Should you run into timeout issues, which has been reported as a common problem, you can run *azdata bdc upgrade* with the additional, optional parameters *controller-timeout* and *component-timeout*. Their value will be in minutes, so if you set them both to 60, it should be more than sufficient.

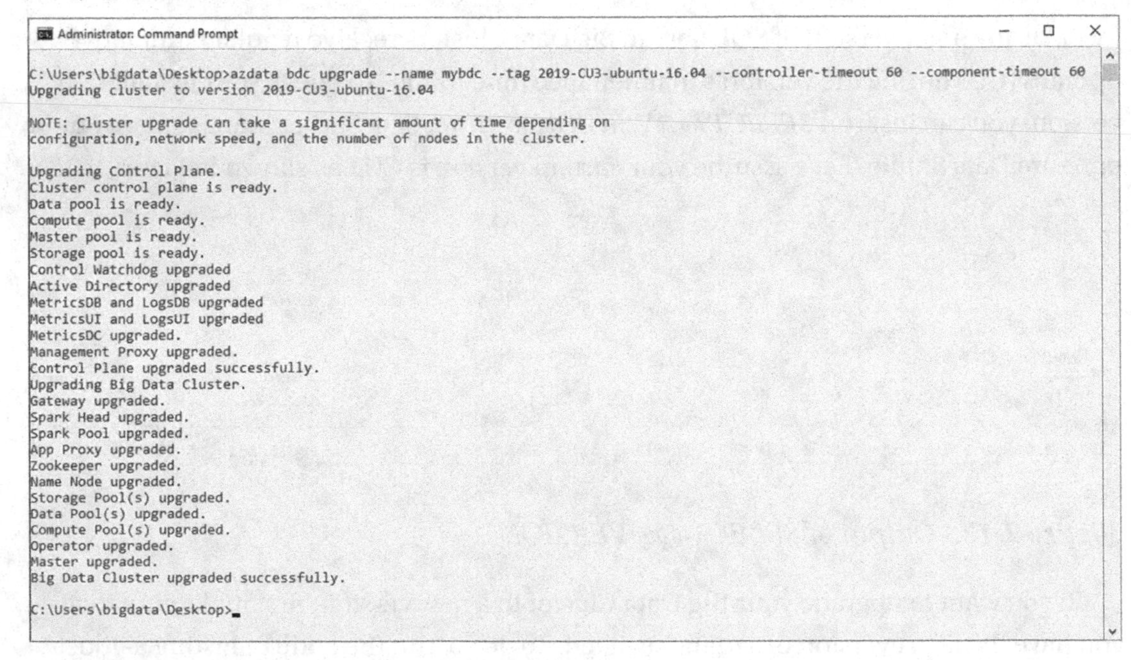


***Figure 9-14.*** *Output of azdata bdc upgrade*

If you now run *SELECT @@VERSION* again, you will see that your Big Data Cluster reflects CU3 as its current version as shown in Figure 9-15.

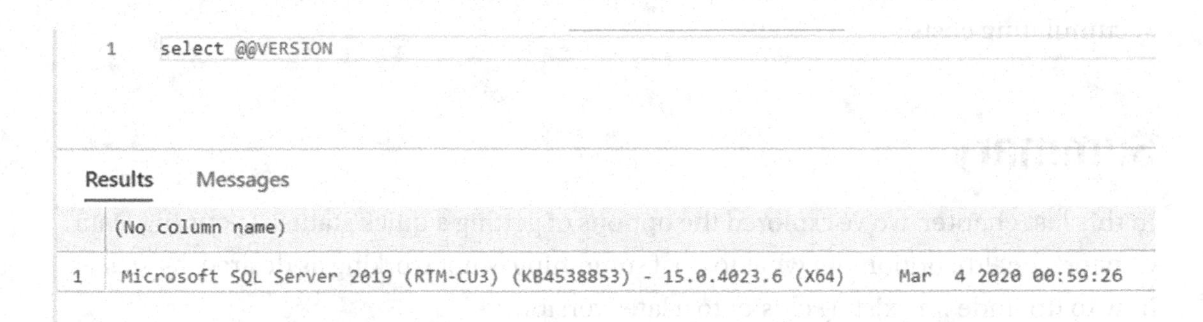


***Figure 9-15.*** *Output of SELECT @@VERSION after upgrade*

# Removing a Big Data Cluster Instance

If you want to delete an instance of your Big Data Cluster, all you need to do is use *azdata* again. You will just need to provide the name of your instance as shown in Listing 9-3 and the cluster components will be deleted.

***Listing 9-3.*** Install script for Chocolatey in PowerShell

```
azdata bdc delete --name <ClusterToBeDeleted>
```

You can follow the progress until the instance is fully removed as shown in Figure 9-16.

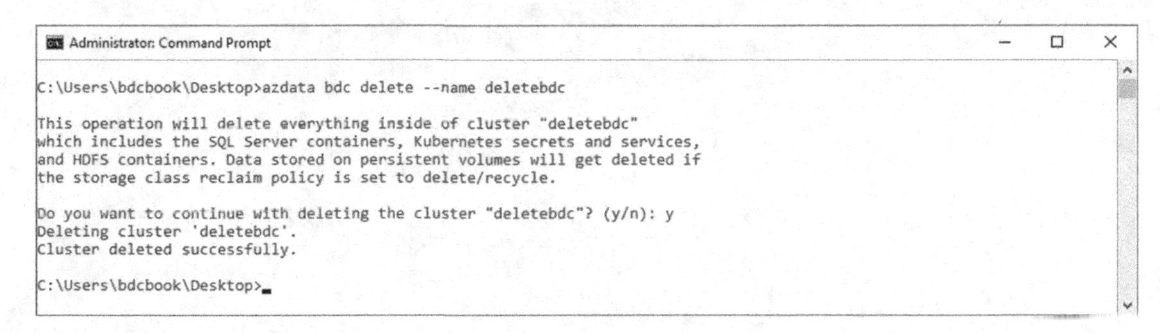


***Figure 9-16.*** *Output of azdata bdc delete*

That's it – your instance is now removed. This will only remove the Big Data Cluster components, so if you deployed to Azure Kubernetes Services, you may want to consider deleting that cluster as well, unless you need it for other applications, to avoid it accumulating costs.

## Summary

In this last chapter, we've explored the options of getting a quick status of your Big Data Cluster's health, options on what to do if something is not working as desired, as well as how to upgrade an existing cluster to a later version.

# Index






B. Weissman and E. van de Laar, *SQL Server Big Data Clusters*,
https://doi.org/10.1007/978-1-4842-5985-6

## D

## J



## K



## L



## M

## N



## O



## P



## Q

## R



## S

## T



## U



## V, W, X, Y, Z

Printed in the United States
By Bookmasters